Hi there, kids! I have some easy instructions for you. To make things easier, it's a good idea to use Crayola Erasable Colored Pencils. That way, if you make a mistake, you can just erase it and try again. I'll be here to help and guide you through your math journey. And if you want to check your answers, just flip the page over to see if you got it right. Remember, math can be fun!

Welcome to our math book for kids ages 4 to 6! This book is filled with fun and exciting activities that will help you learn all about numbers and counting.

Each section of the book is designed to help you build your math skills step-by-step. The first section is all about tracing numbers from 1 to 10. You'll learn how to write each number by following the dotted lines.

Next, you'll move on to counting and coloring objects. You'll practice counting different types of objects, like apples, cars, and animals. Then, you'll get to color them in however you like!

After that, it's time to count and match the numbers. You'll practice matching numbers with the right amount of objects, like matching the number 3 with a picture of three Tigers.

In the "I Spy" section, you'll count the objects on the page and then write the correct number. It's like a fun "I Spy" game that helps you learn to count.

In the next section, you'll practice adding numbers together. You'll count how many objects there are, add them together, and then write the total number.

Next, you'll get to do some dot-to-dot exercises from 1 to 10. You'll connect the dots to make a picture, and then color it in.

Finally, you'll learn about subtraction. You'll cross out the correct amount of objects and then write the number left over.

Remember to have fun with each activity and take your time. You can do them in any order you like, and if you get stuck, don't worry! Just ask an adult for help.

Let's get started and have fun learning math!

Welcome to the number tracing pages! This is a fun activity that will help you learn how to write the numbers from 1 to 10. To get started, grab a pencil and get ready to trace! Take your time and try your best. Remember, it's okay to make mistakes. Have fun!

ONE

Trace the numbers

1
1
1

Write the number 1

TWO

Trace the numbers

2

2

2

Write the number 2

THREE 3

Trace the numbers

3
3
3

Write the number 3

FOUR

Trace the numbers

4   4   4   4   4   4   4

4   4   4   4   4   4   4

4   4   4   4   4   4   4

Write the number 4

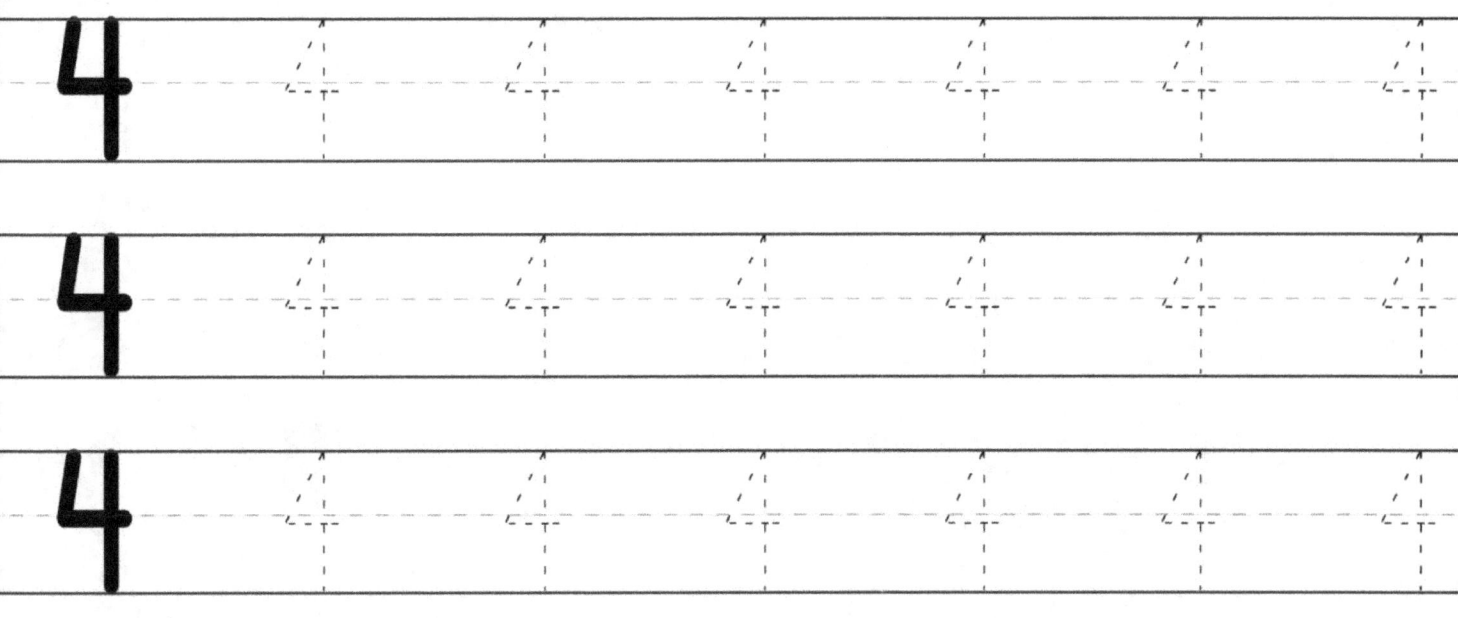

Five

Trace the numbers

5

5

5

Write the number 5

Six

Trace the numbers

6   6   6   6   6   6   6

6   6   6   6   6   6   6

6   6   6   6   6   6   6

Write the number 6

Seven 7

Trace the numbers

7

7

7

Write the number 7

Eight 8

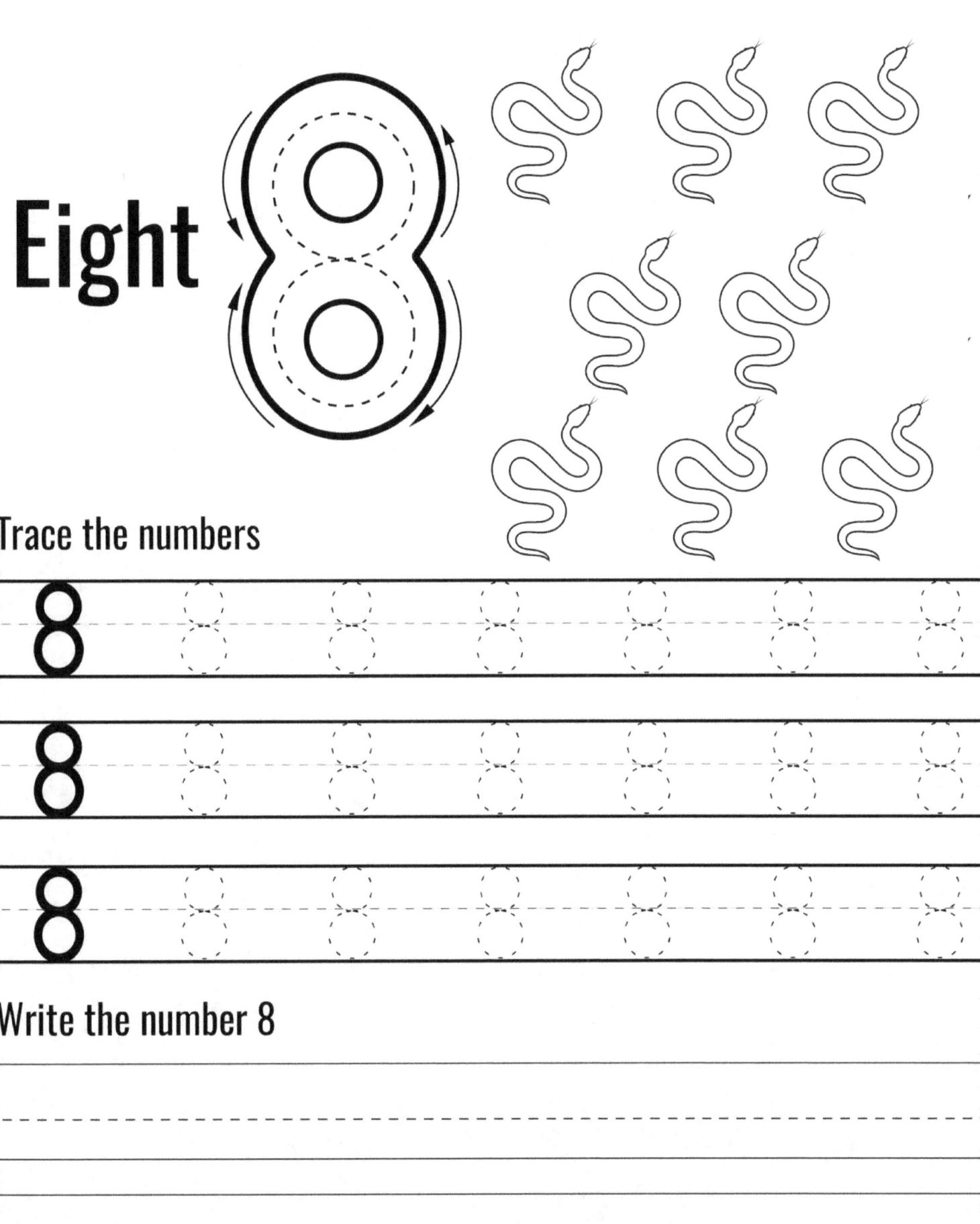

Trace the numbers

8

8

8

Write the number 8

Nine

Trace the numbers

9   9   9   9   9   9   9

9   9   9   9   9   9   9

9   9   9   9   9   9   9

Write the number 9

Ten

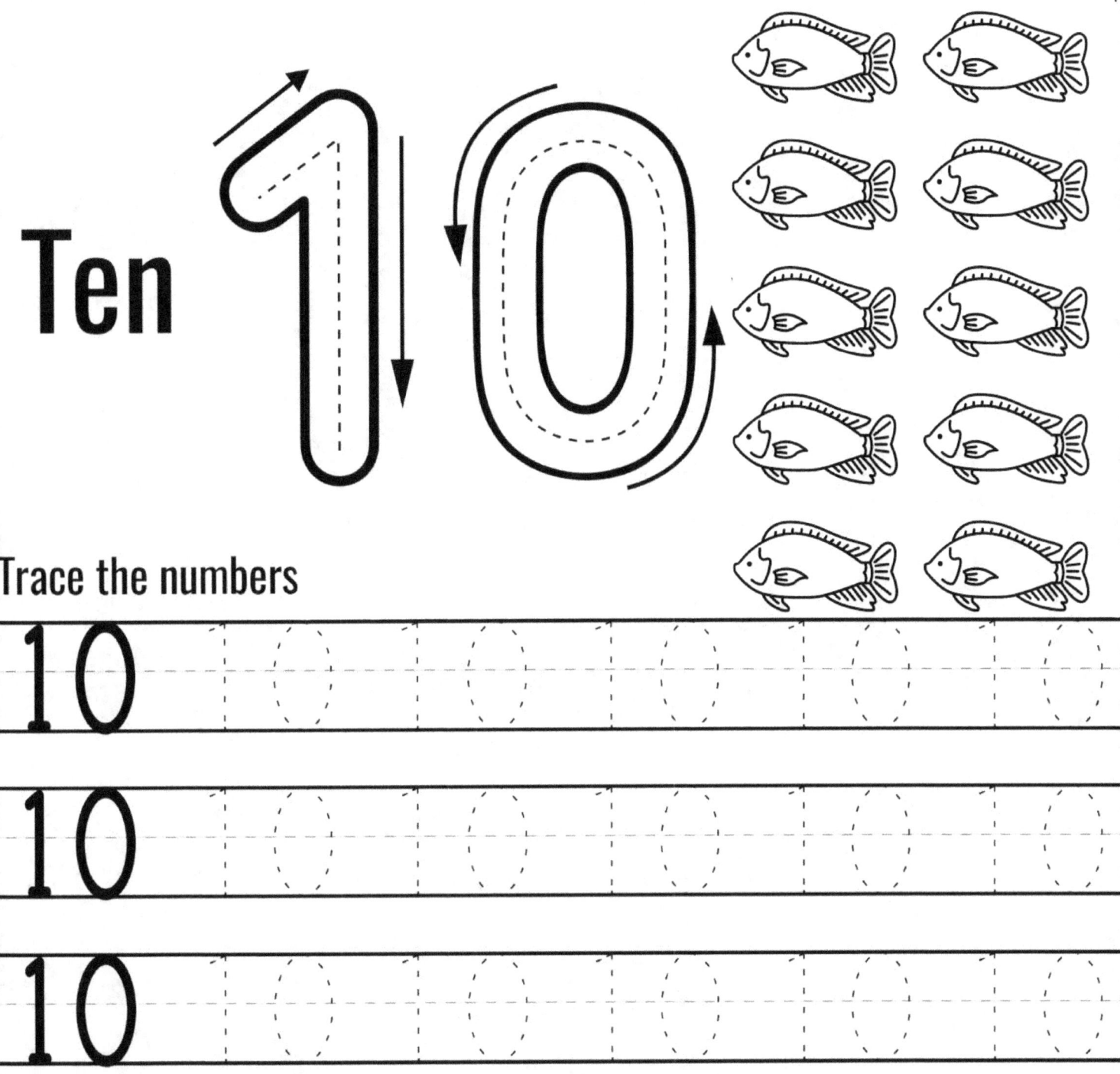

Trace the numbers

10

10

10

Write the number 10

Count and color with us! Grab some crayons and match the number to the amount of objects. Use your favorite colors and have fun!

Let's count the objects and color the correct number.

Congrats! You did it! Here's the answer for you to compare with what you got.

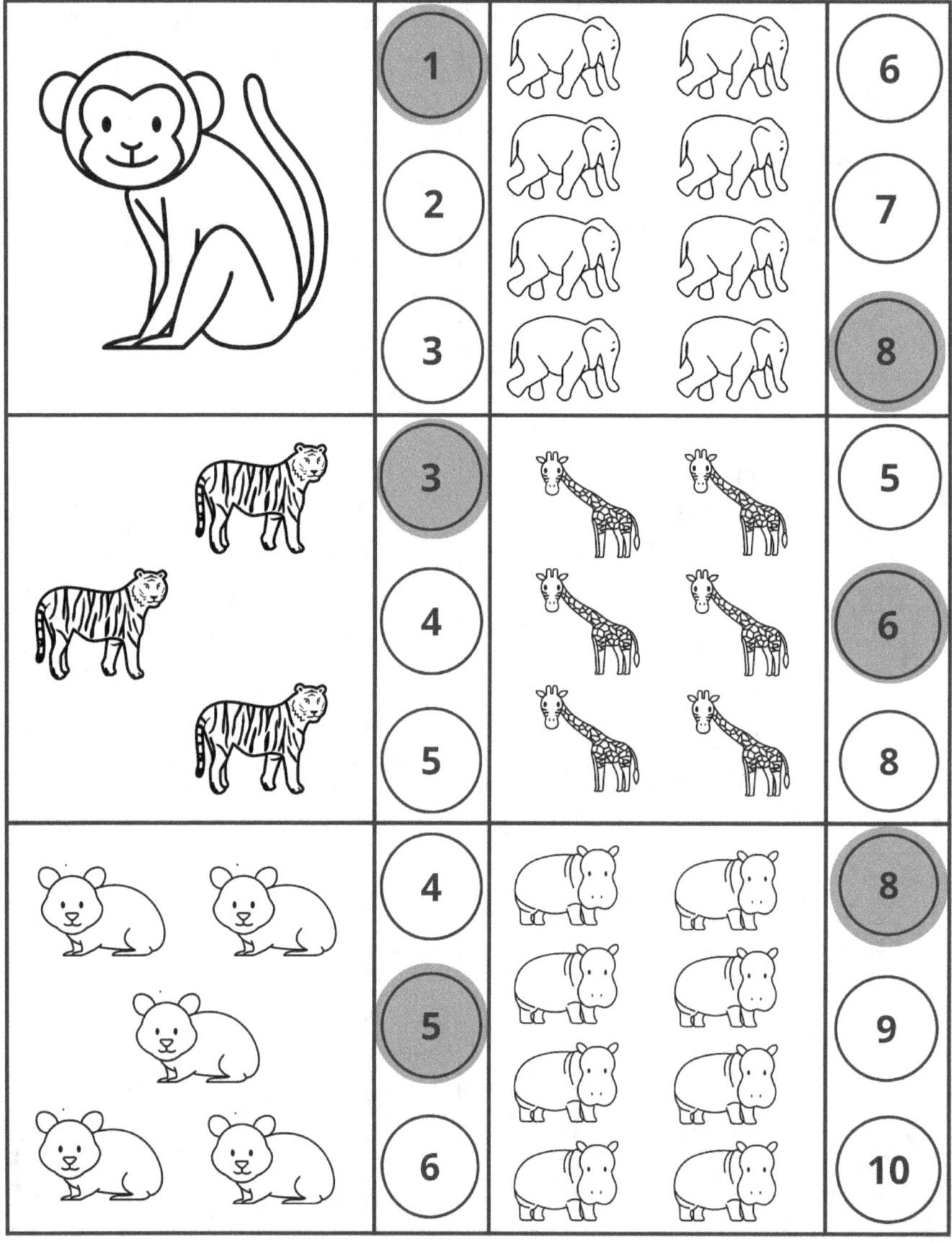

Let's count the objects and color the correct number.

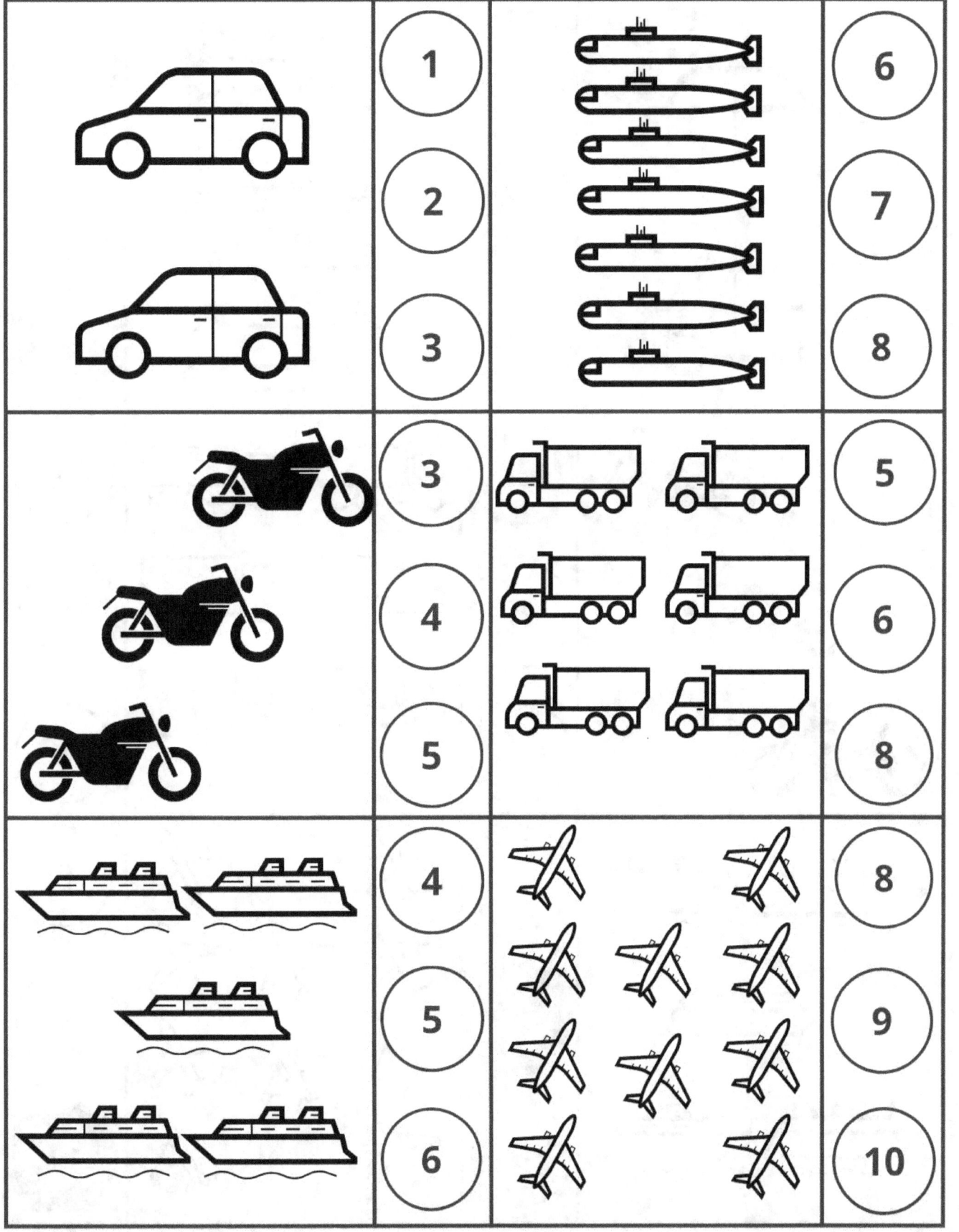

Congrats! You did it! Here's the answer for you to compare with what you got.

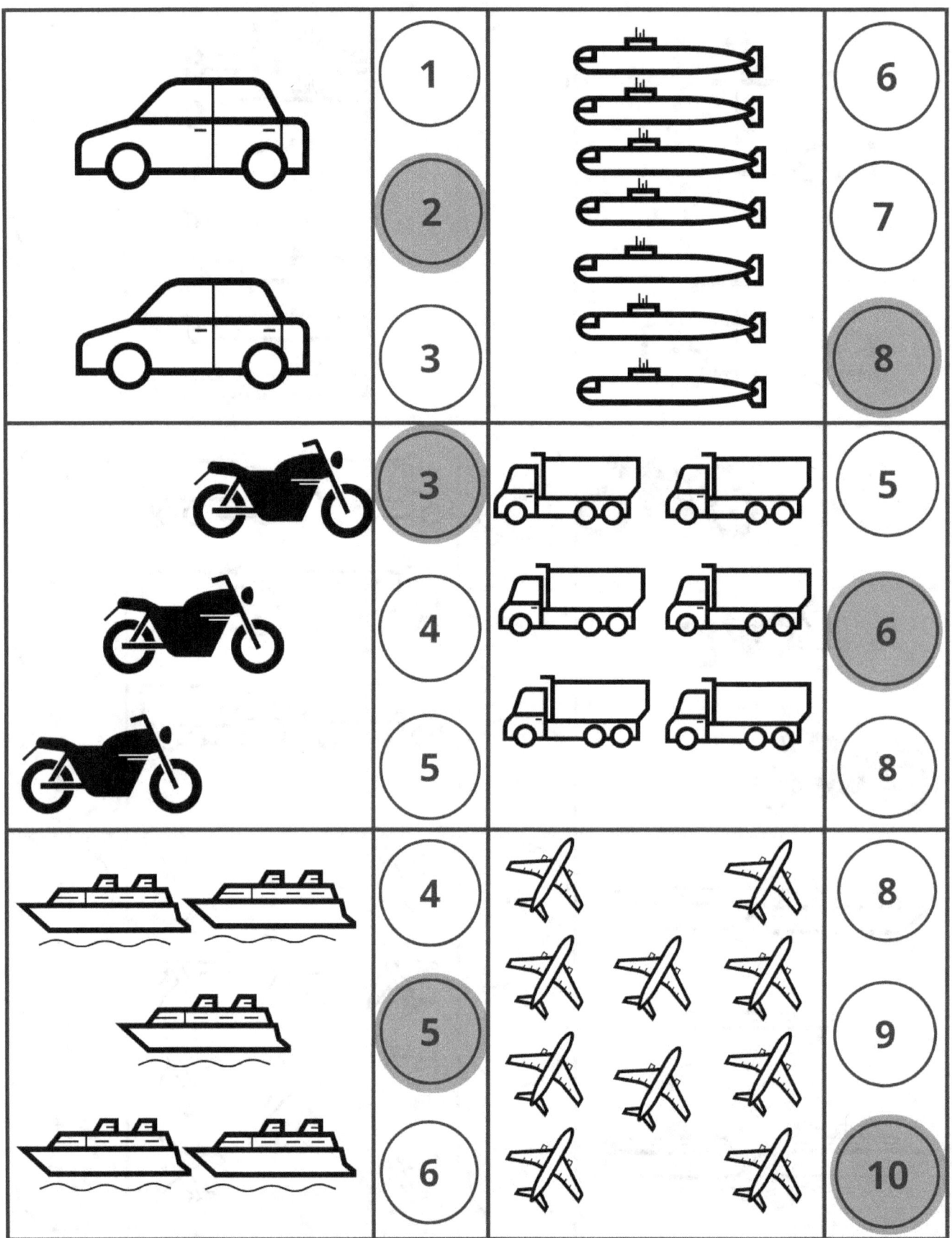

Let's count the objects and color the correct number.

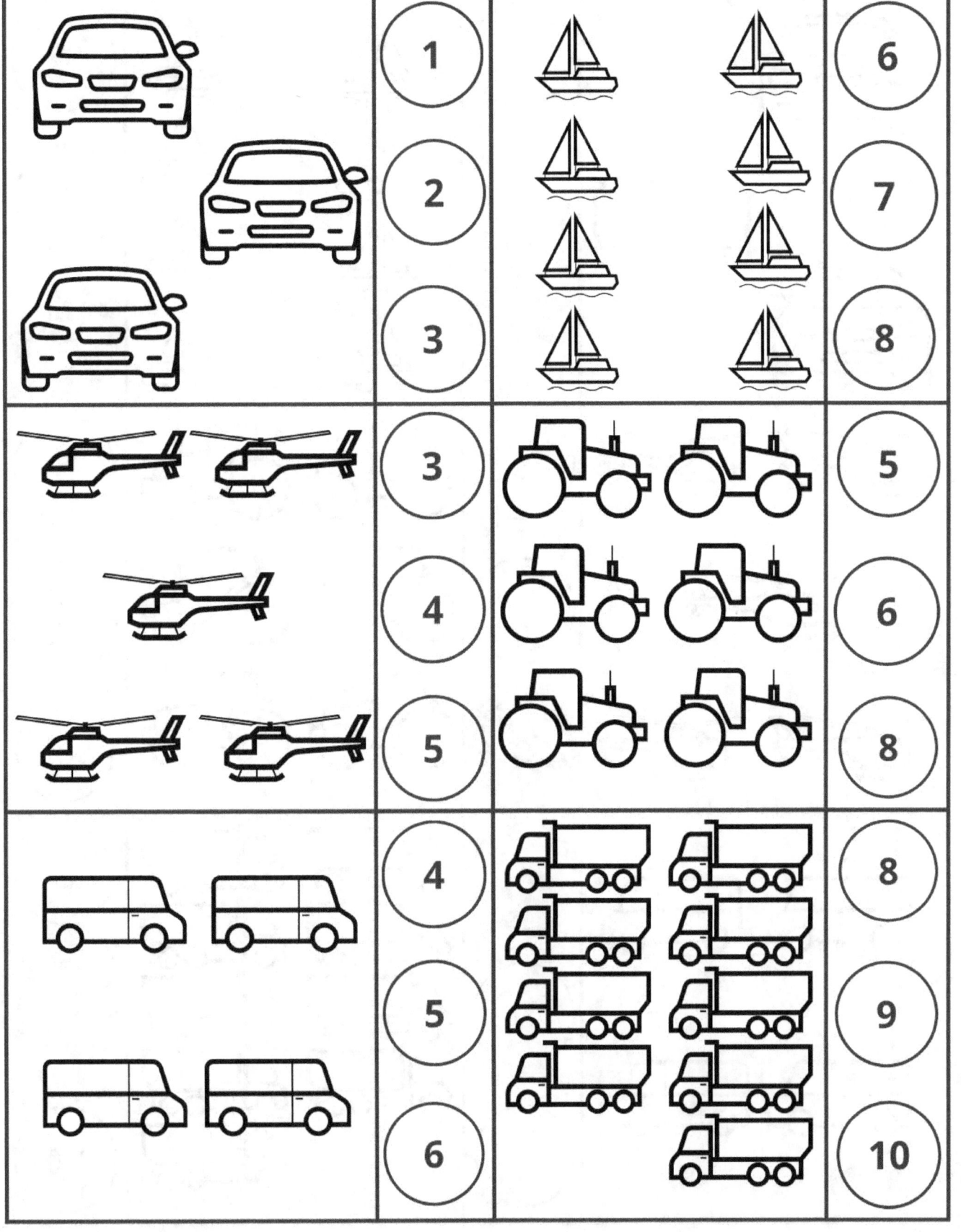

Congrats! You did it! Here's the answer for you to compare with what you got.

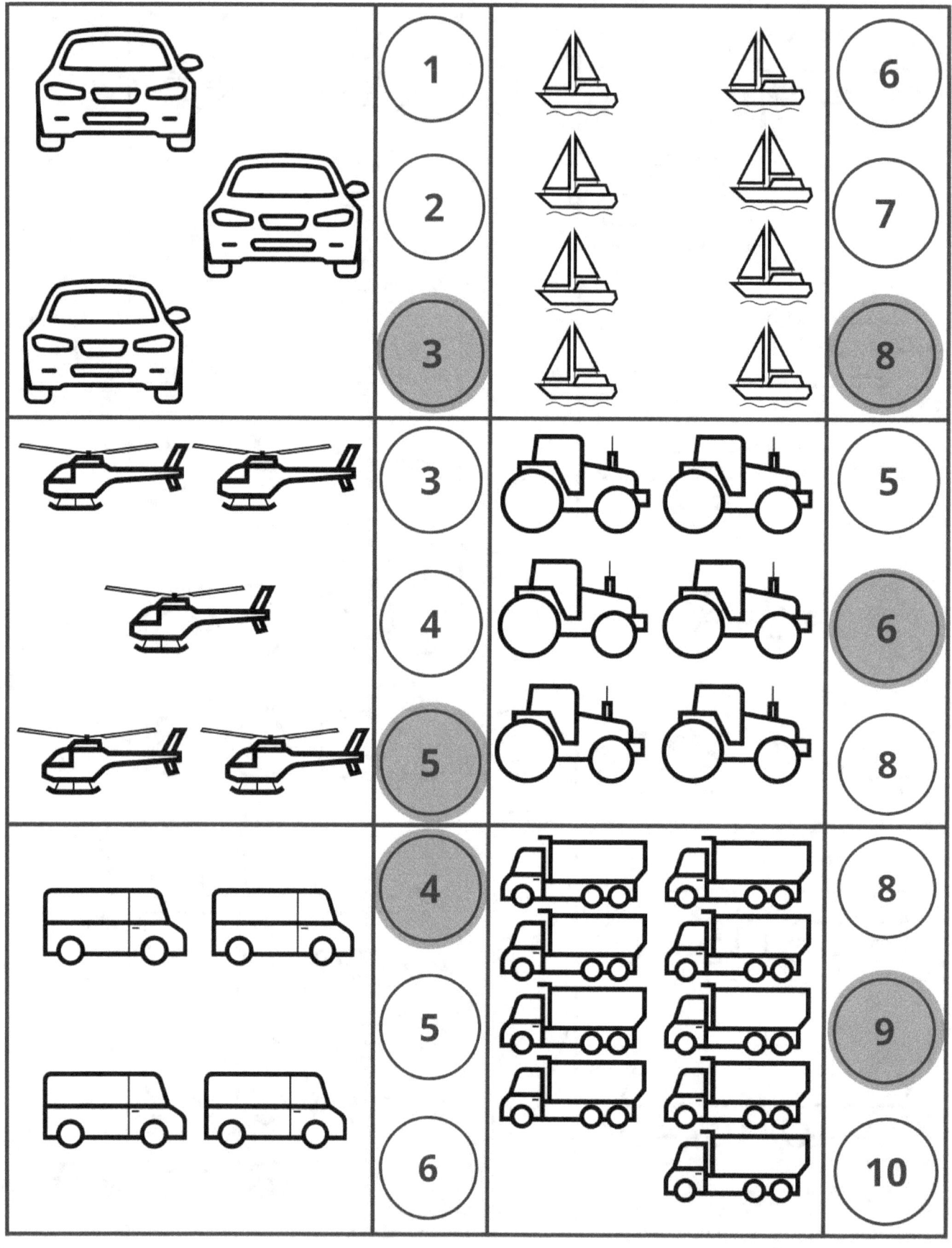

Let's count the objects and color the correct number.

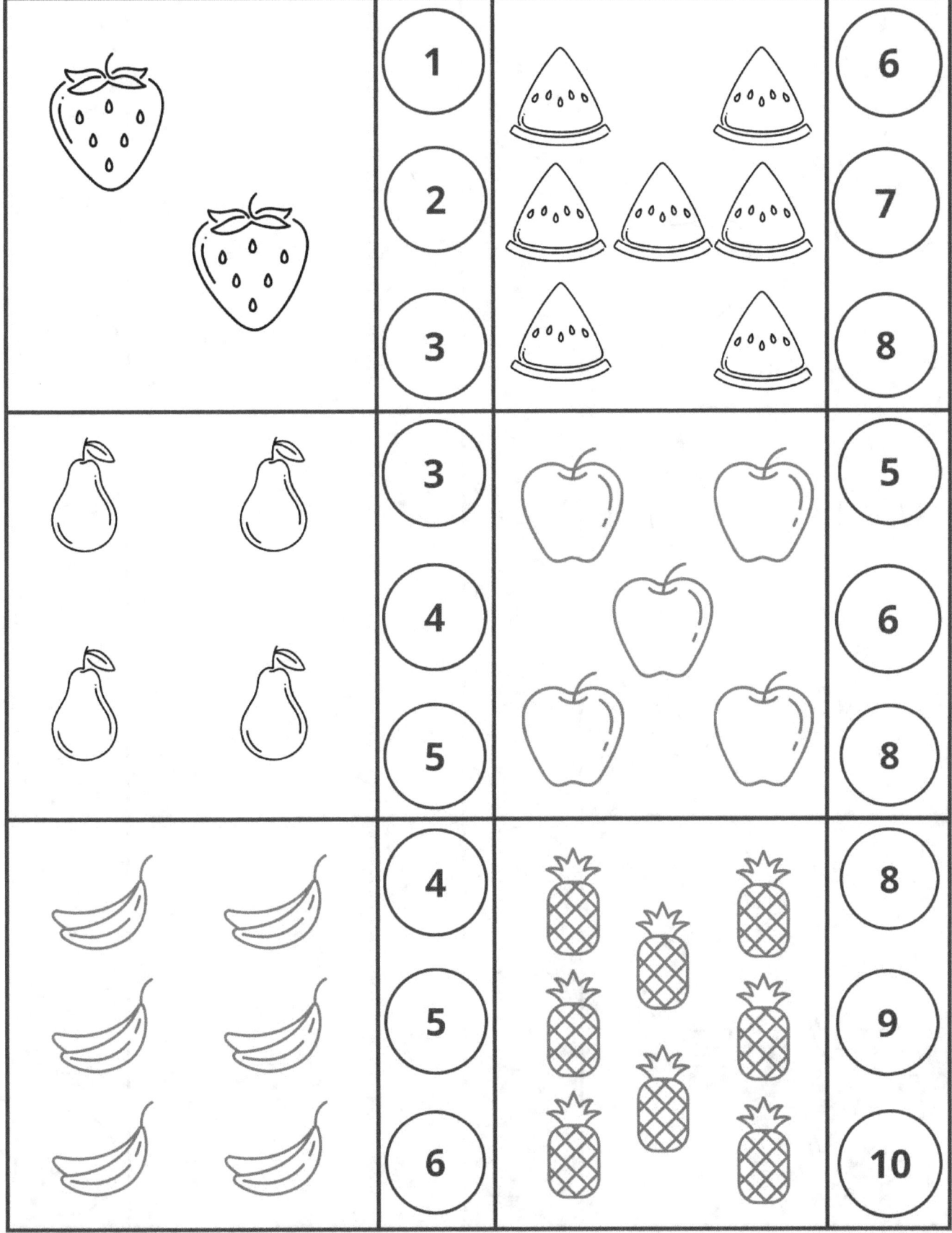

Congrats! You did it! Here's the answer for you to compare with what you got.

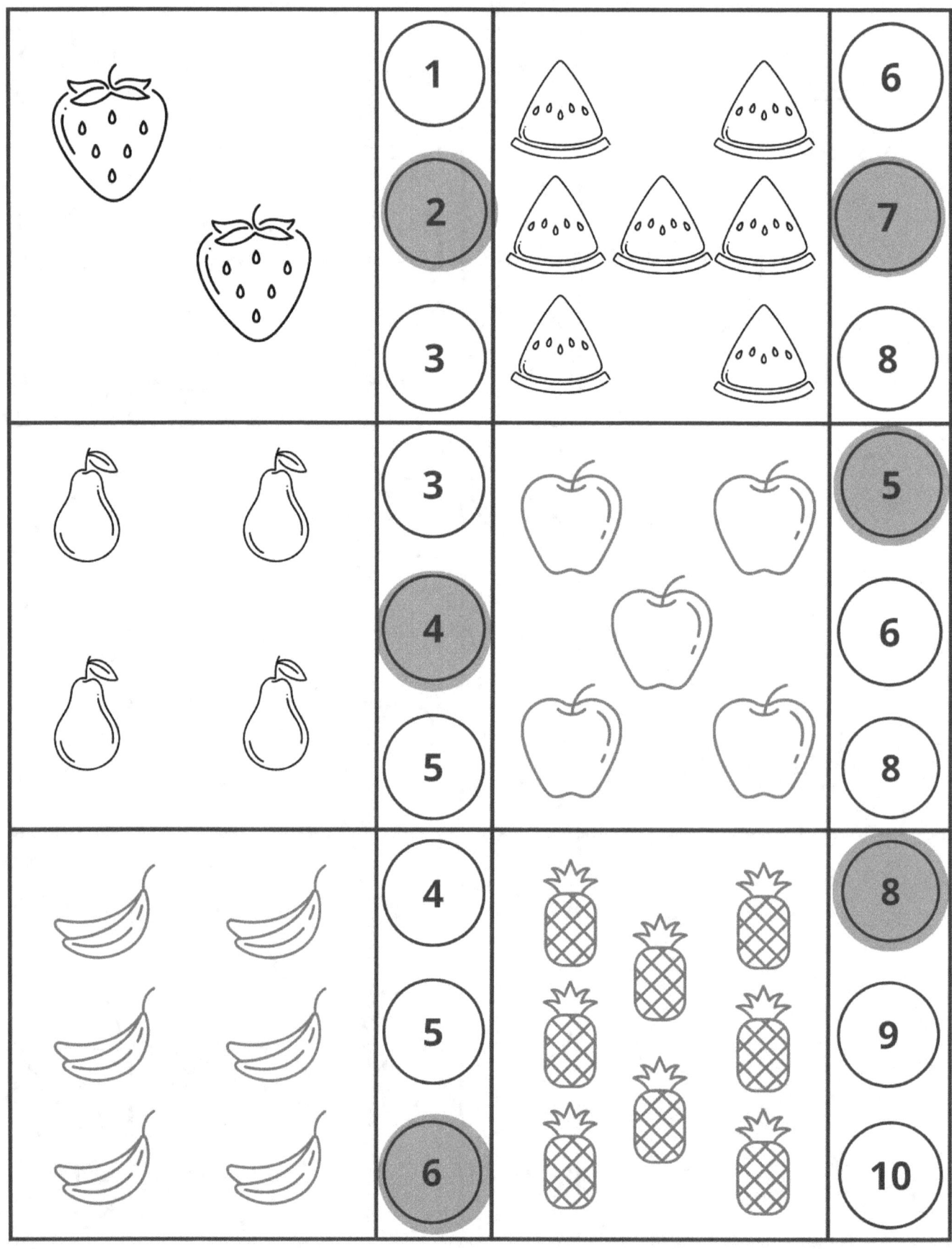

Let's count the objects and color the correct number.

Congrats! You did it! Here's the answer for you to compare with what you got.

Count and match with us!
Look at the pictures, count
the objects, and draw a line
to connect them to the
correct number. Use your
favorite colors and have fun!

Let's count the objects and match the numbers

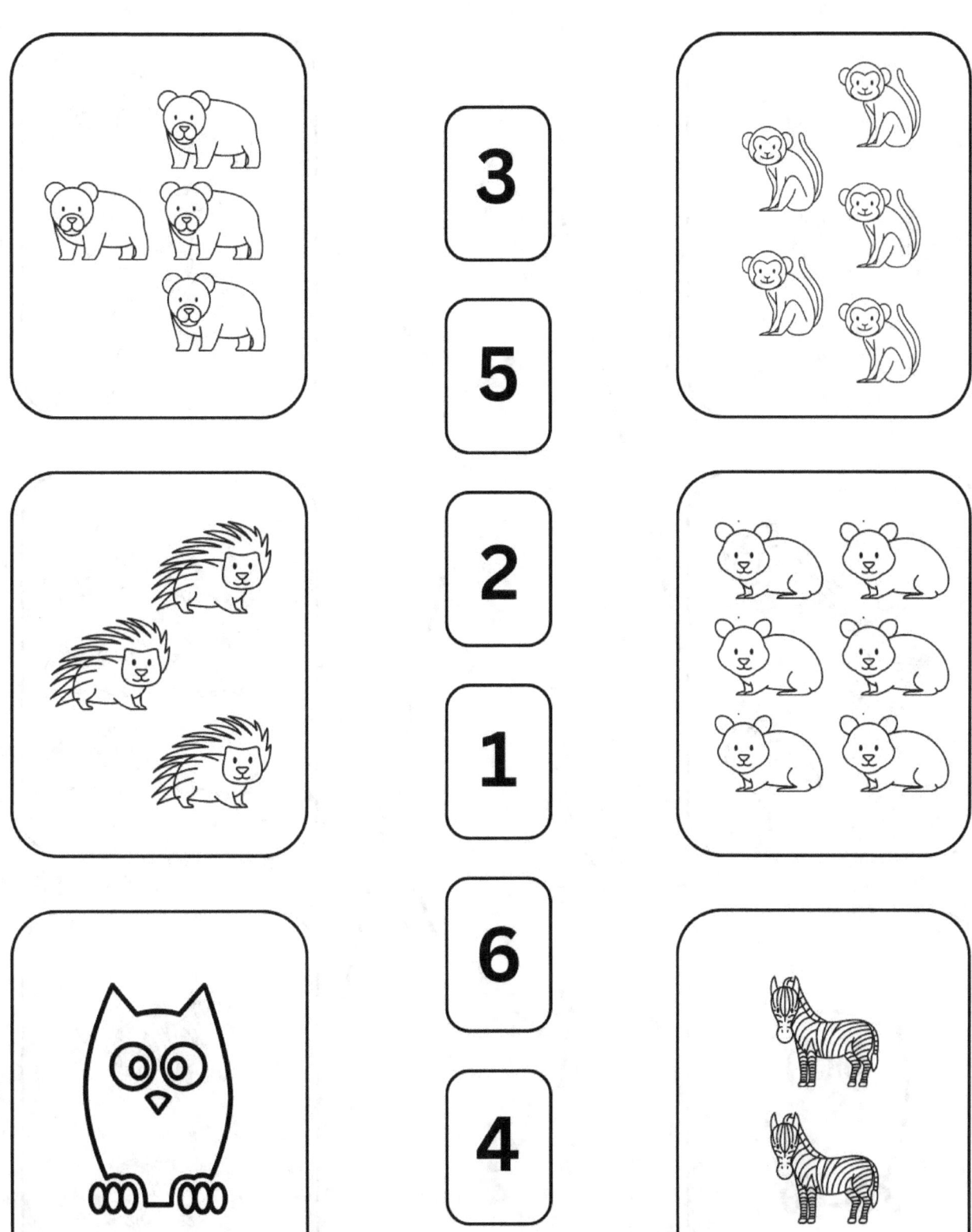

Congrats! You did it! Here's the answer for you to compare with what you got.

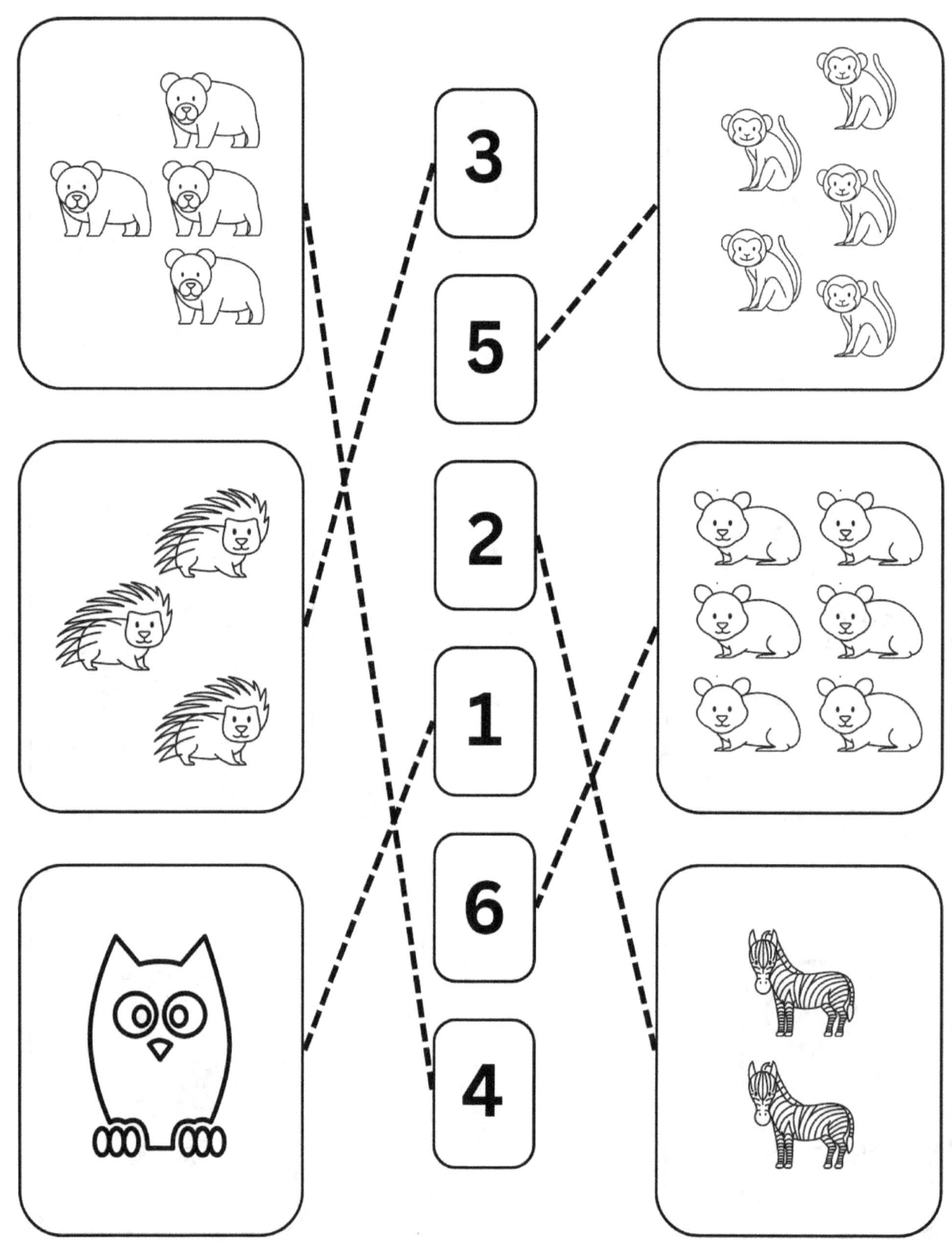

Let's count the objects and match the numbers

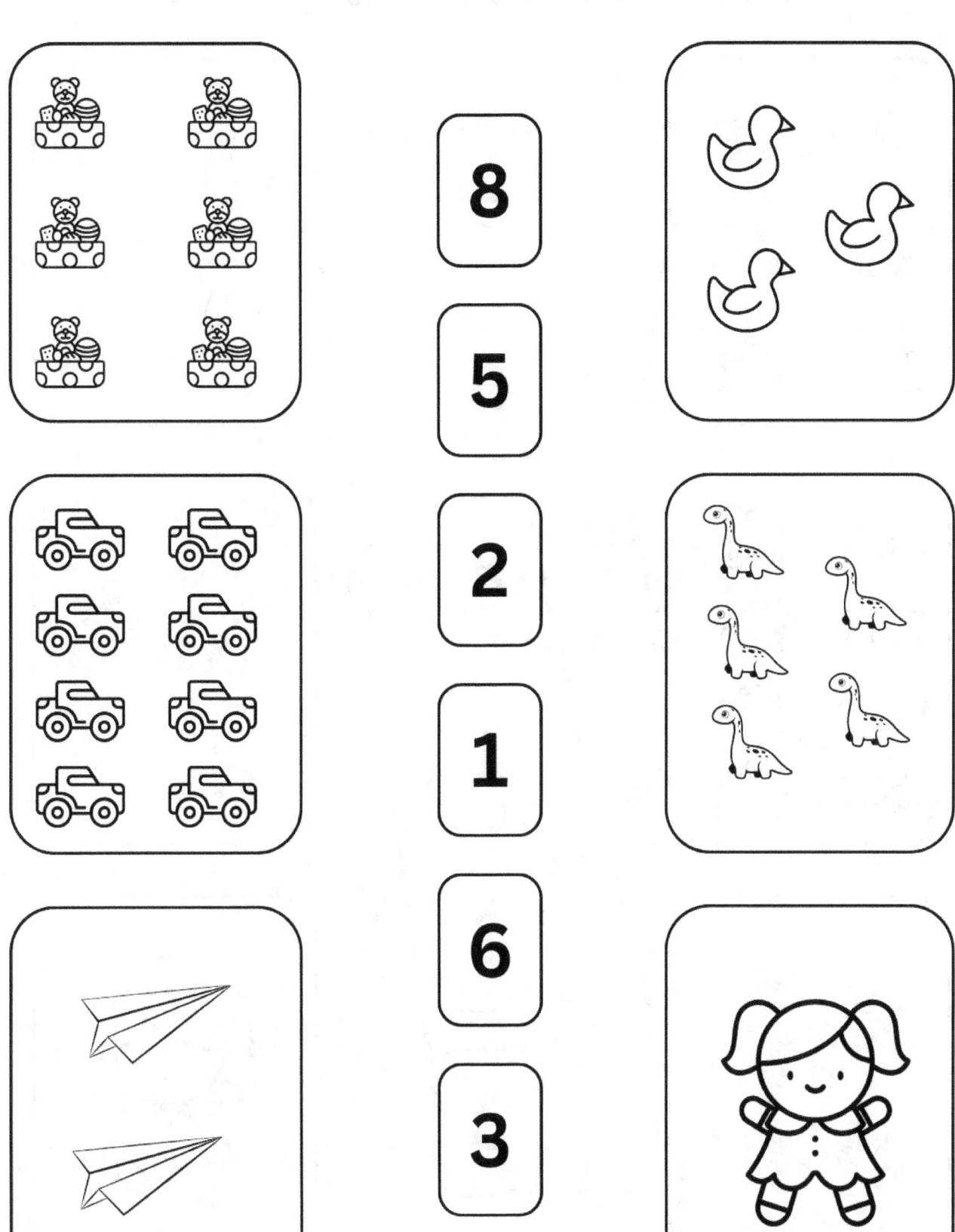

Congrats! You did it! Here's the answer for you to compare with what you got.

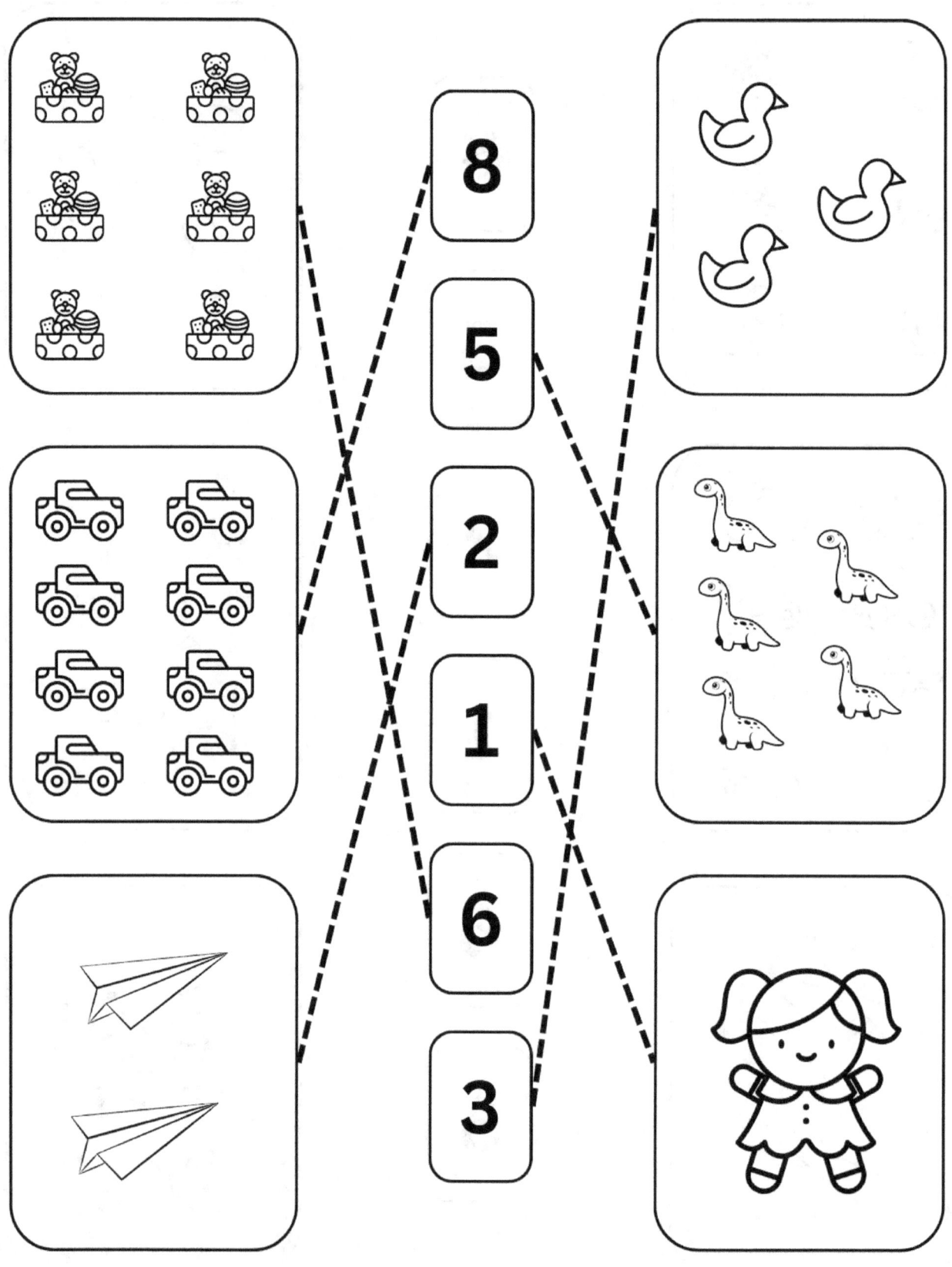

Let's count the objects and match the numbers

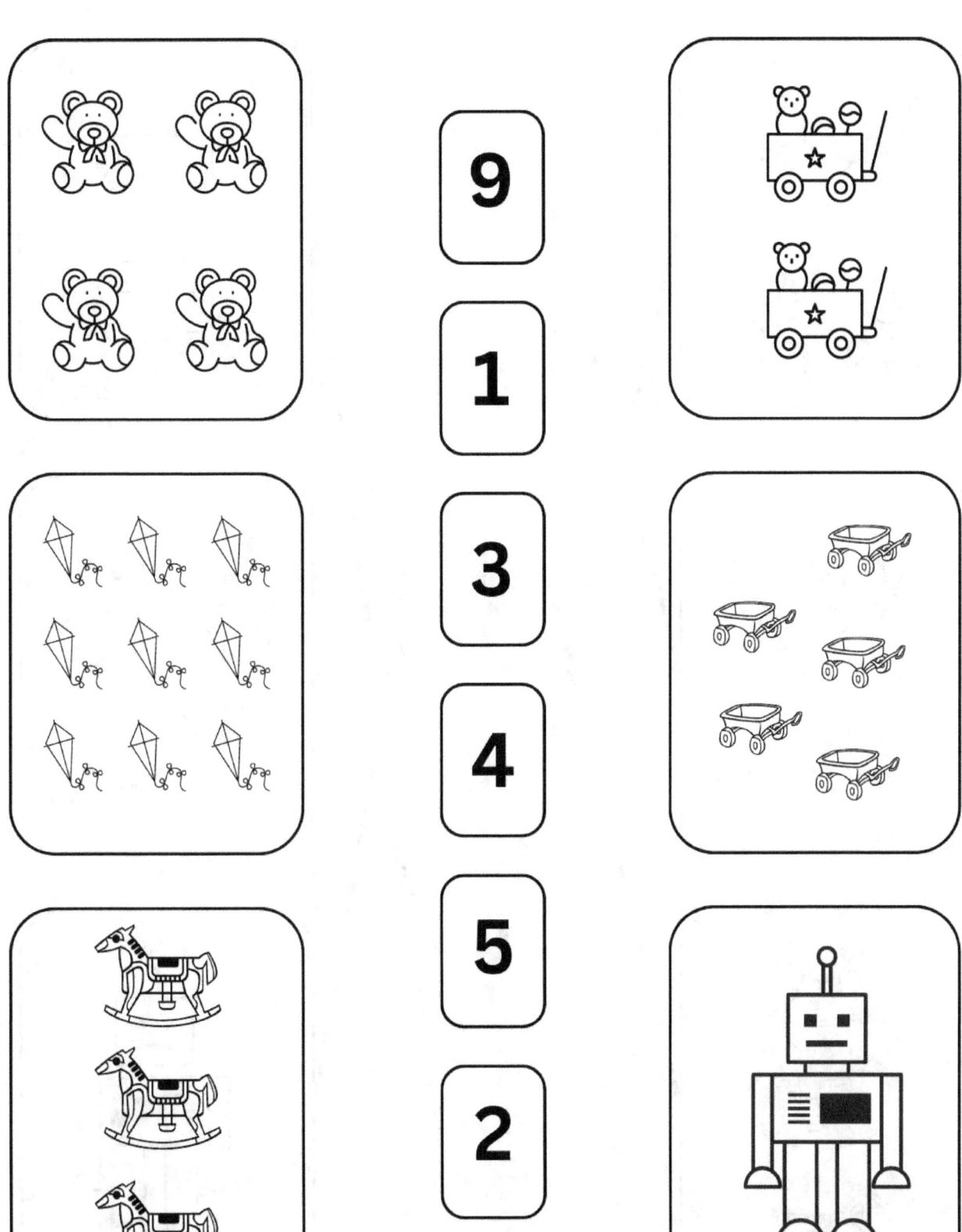

Congrats! You did it! Here's the answer for you to compare with what you got.

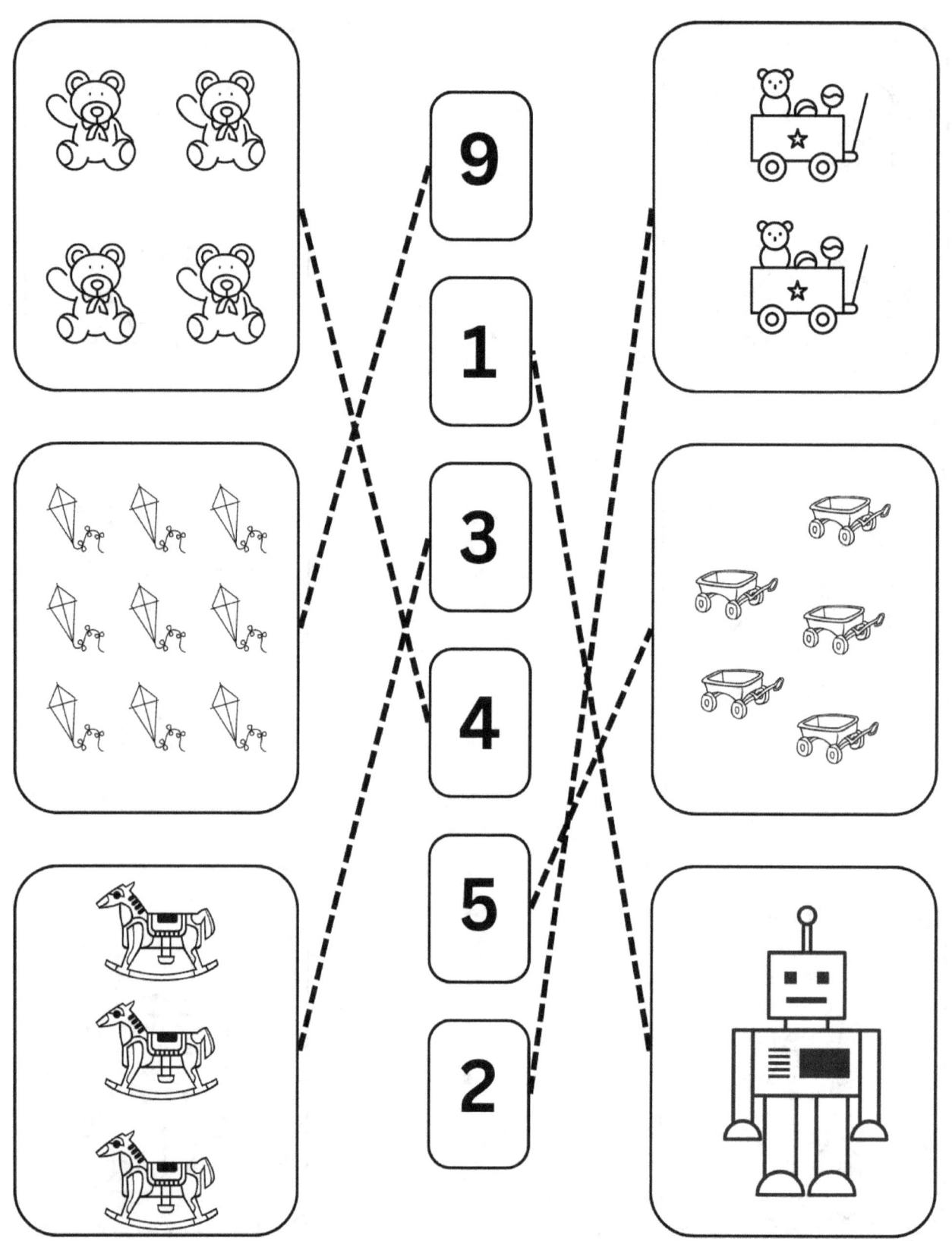

Let's count the objects and match the numbers

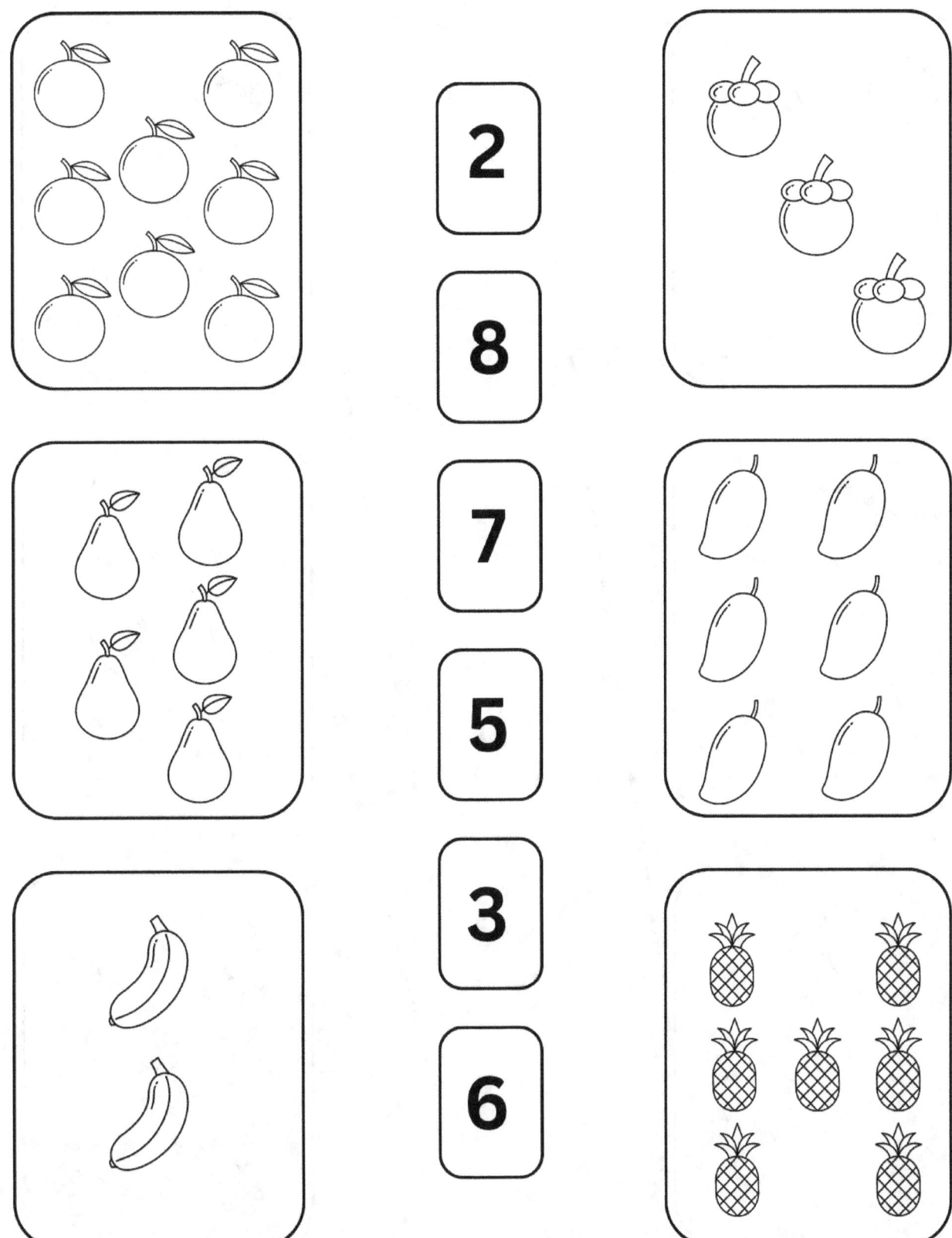

Congrats! You did it! Here's the answer for you to compare with what you got.

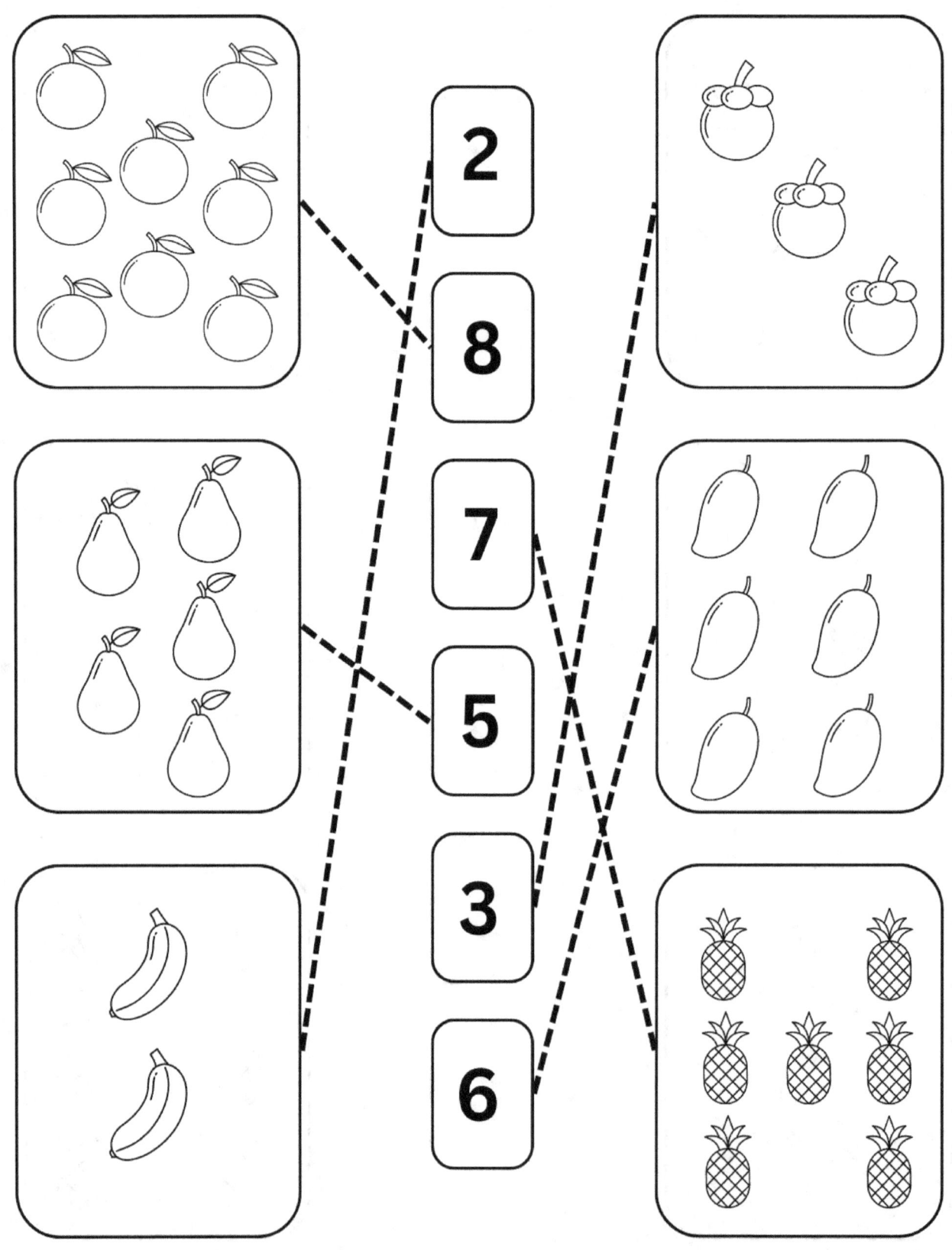

Let's count the objects and match the numbers

Congrats! You did it! Here's the answer for you to compare with what you got.

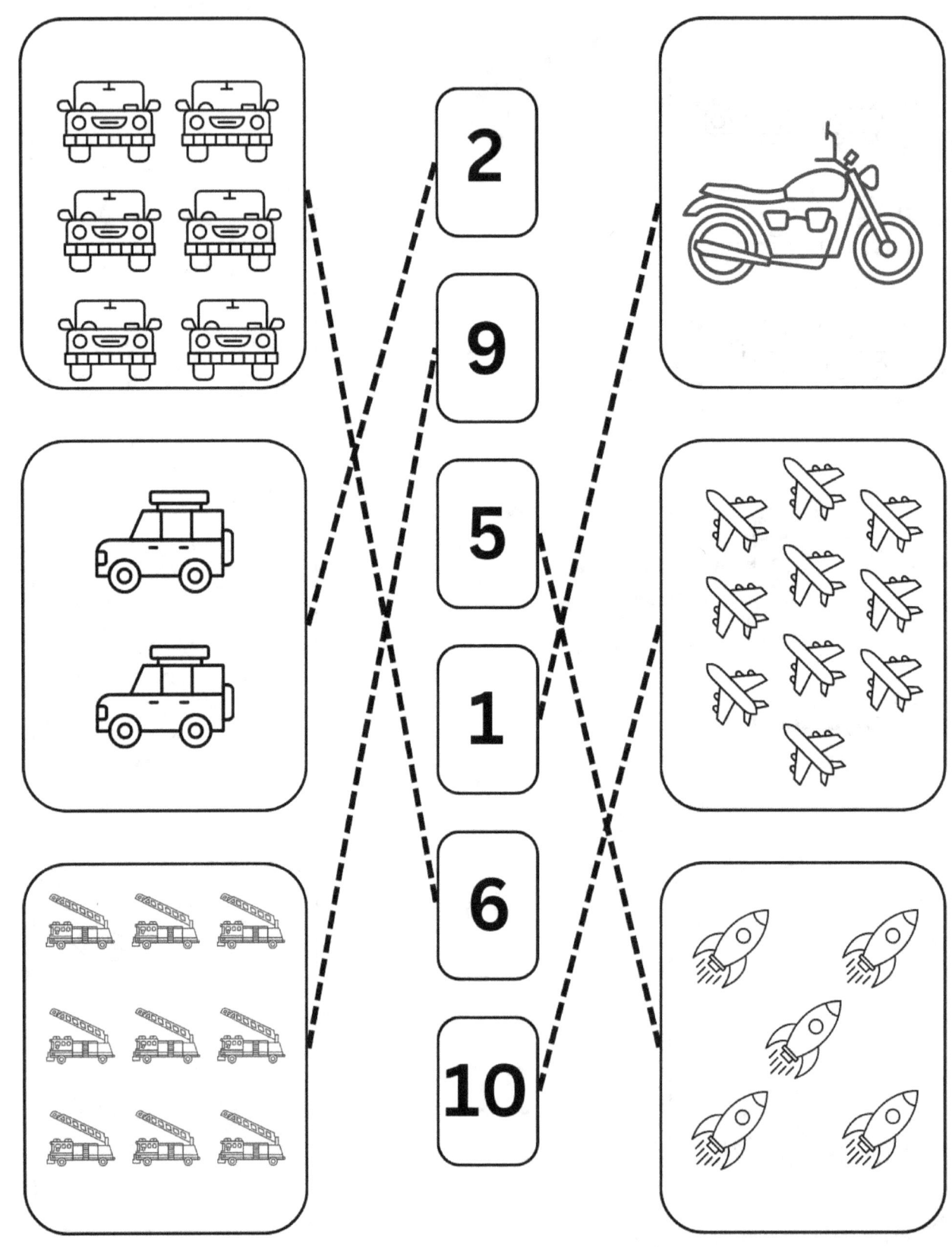

Let's count the objects and match the numbers

Congrats! You did it! Here's the answer for you to compare with what you got.

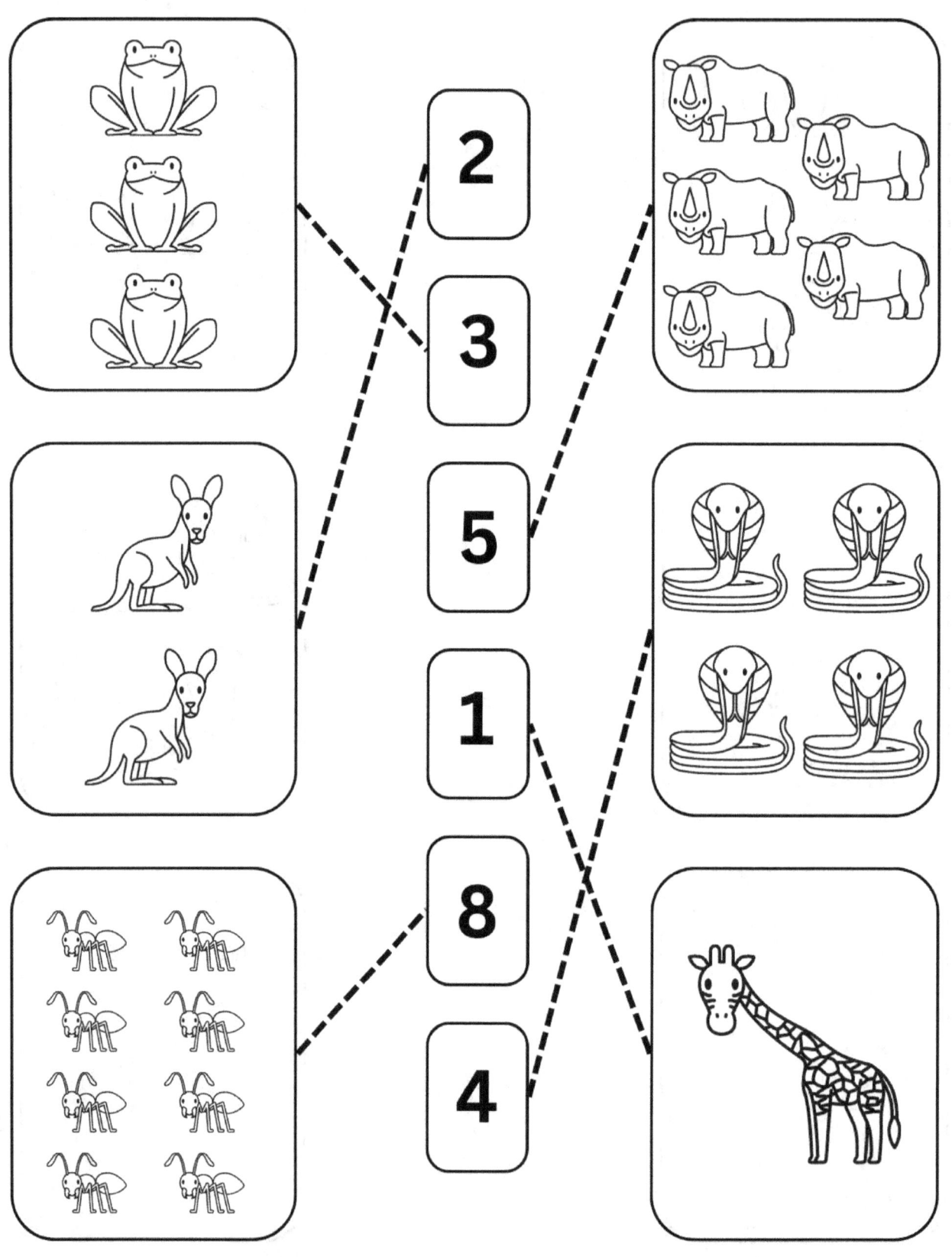

Let's play I Spy Count and Write! Count how many objects you spy and write the number next to it. Use your favorite colors and have fun!

I SPY
How many do you see?

 =    =   =

 =    =    =

Congrats! You did it! Here's the answer for you to compare with what you got.

 = 3  =10  = 2

 = 4  = 4  = 5

I SPY
How many do you see?

Congrats! You did it! Here's the answer for you to compare with what you got.

I SPY
How many do you see?

 =    =

 =   =   =

=   =   =

Congrats! You did it! Here's the answer for you to compare with what you got.

 = 6   = 4   = 5

= 2   = 3   = 1

I SPY
How many do you see?

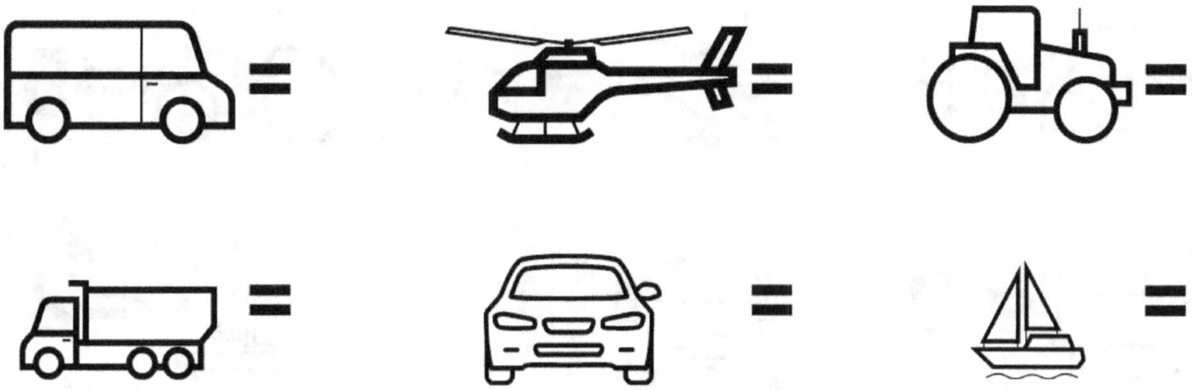

 =    =

I SPY
How many do you see?

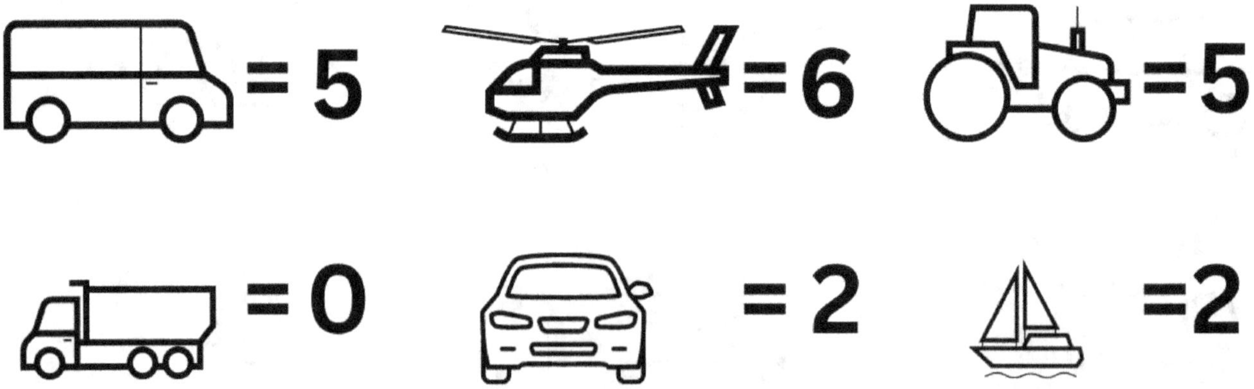

 = 5

 = 6

= 5

= 0

 = 2

 = 2

I SPY
How many do you see?

Congrats! You did it! Here's the answer for you to compare with what you got.

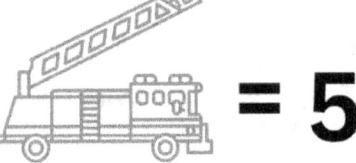 = 5

 =1

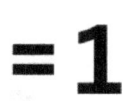 = 2

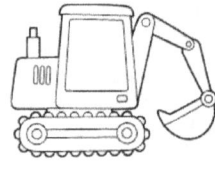 = 2

 =3

 = 5

I SPY
How many do you see?

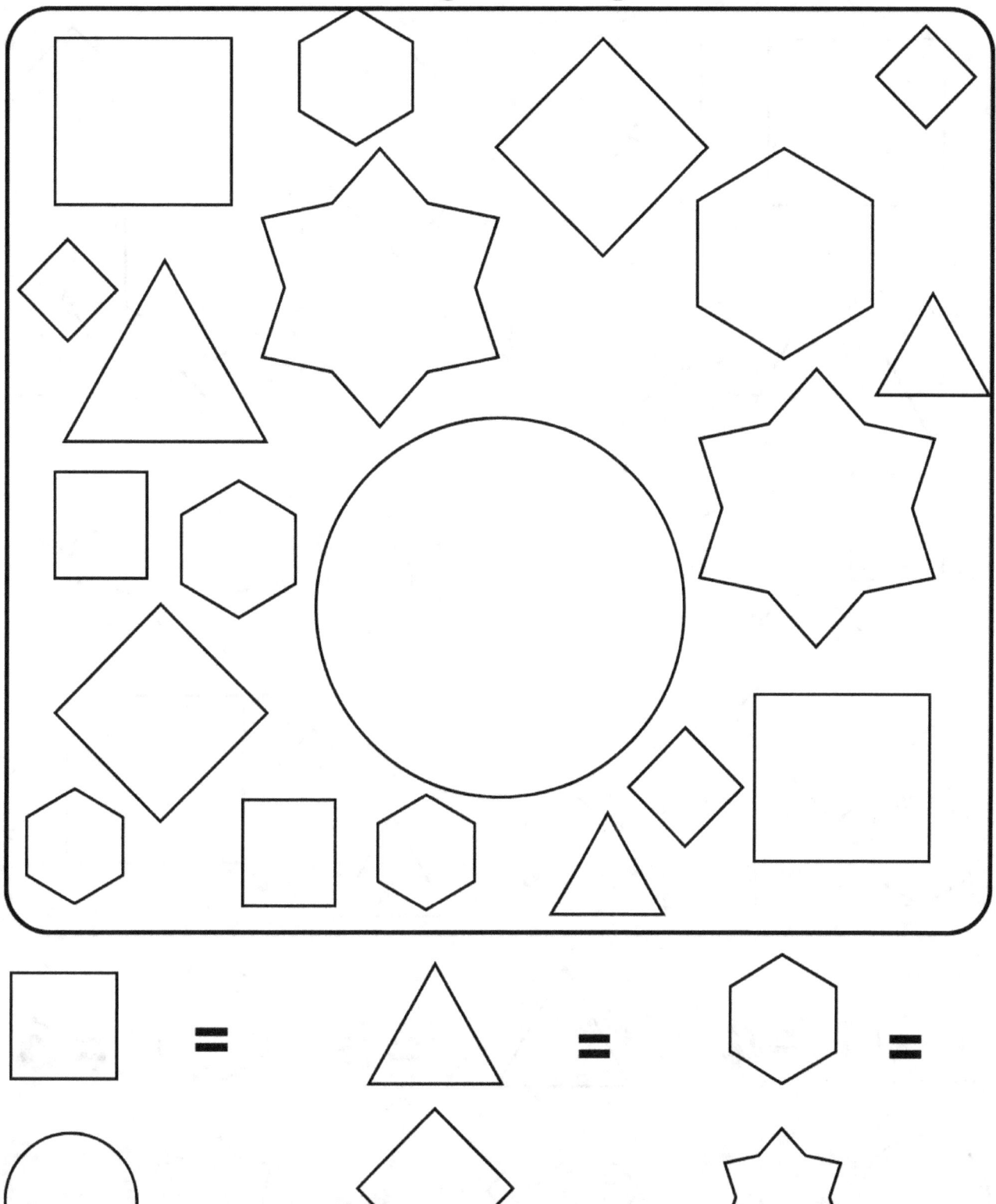

Congrats! You did it! Here's the answer for you to compare with what you got.

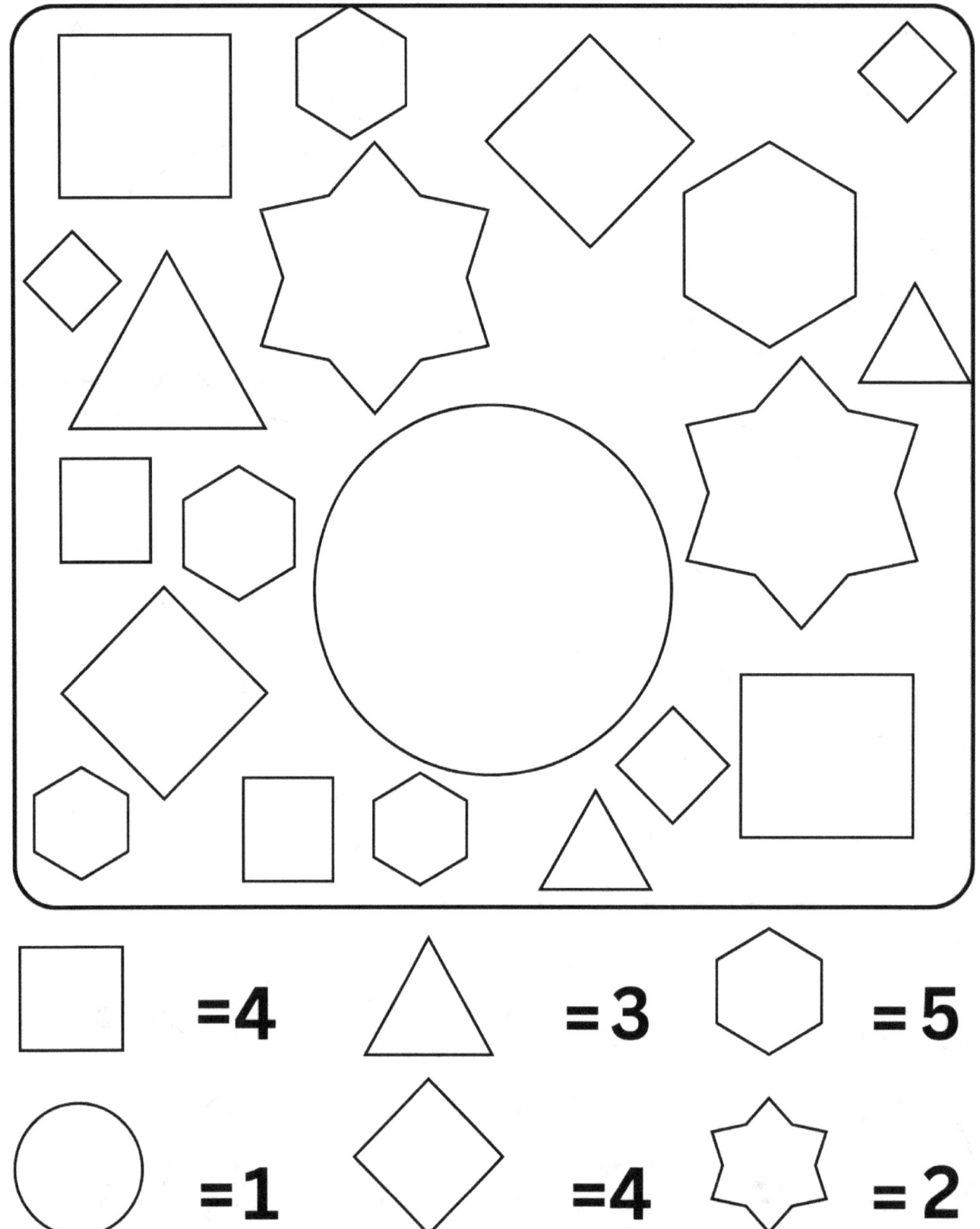

□ =4 △ =3 ⬡ =5

○ =1 ◇ =4 ✶ =2

Count, add, and write with us! Count the objects, add them together, and write the total in the last box. Use your favorite colors and have fun!

Count, add and write the sum in the box.

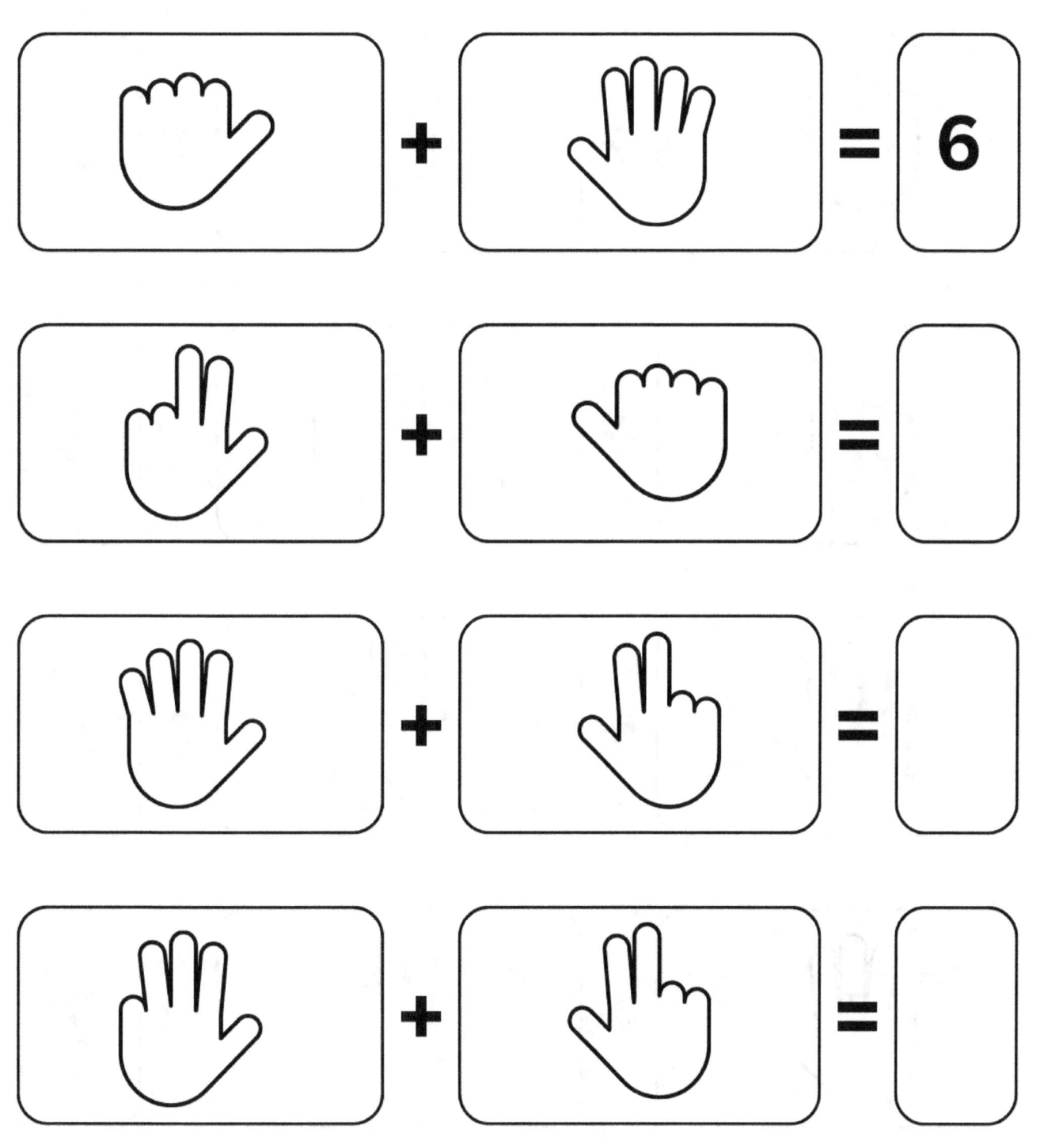

Congrats! You did it! Here's the answer for you to compare with what you got.

+  = 6

+  = 4

+  = 8

+  = 7

Count, add and write the sum in the box.

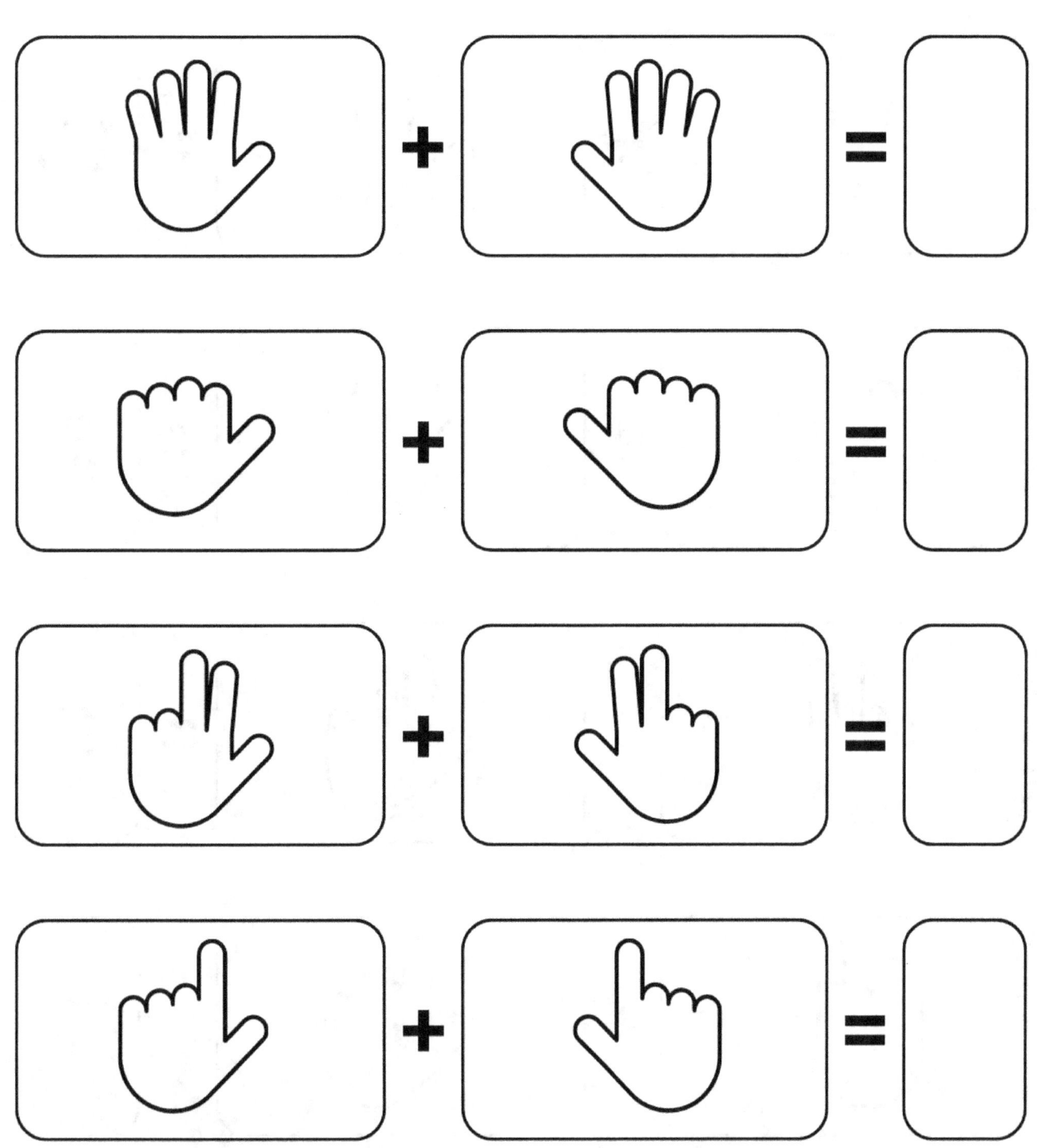

Congrats! You did it! Here's the answer for you to compare with what you got.

+ = 10

+  = 2

+  = 6

+  = 4

Count, add and write the sum in the box.

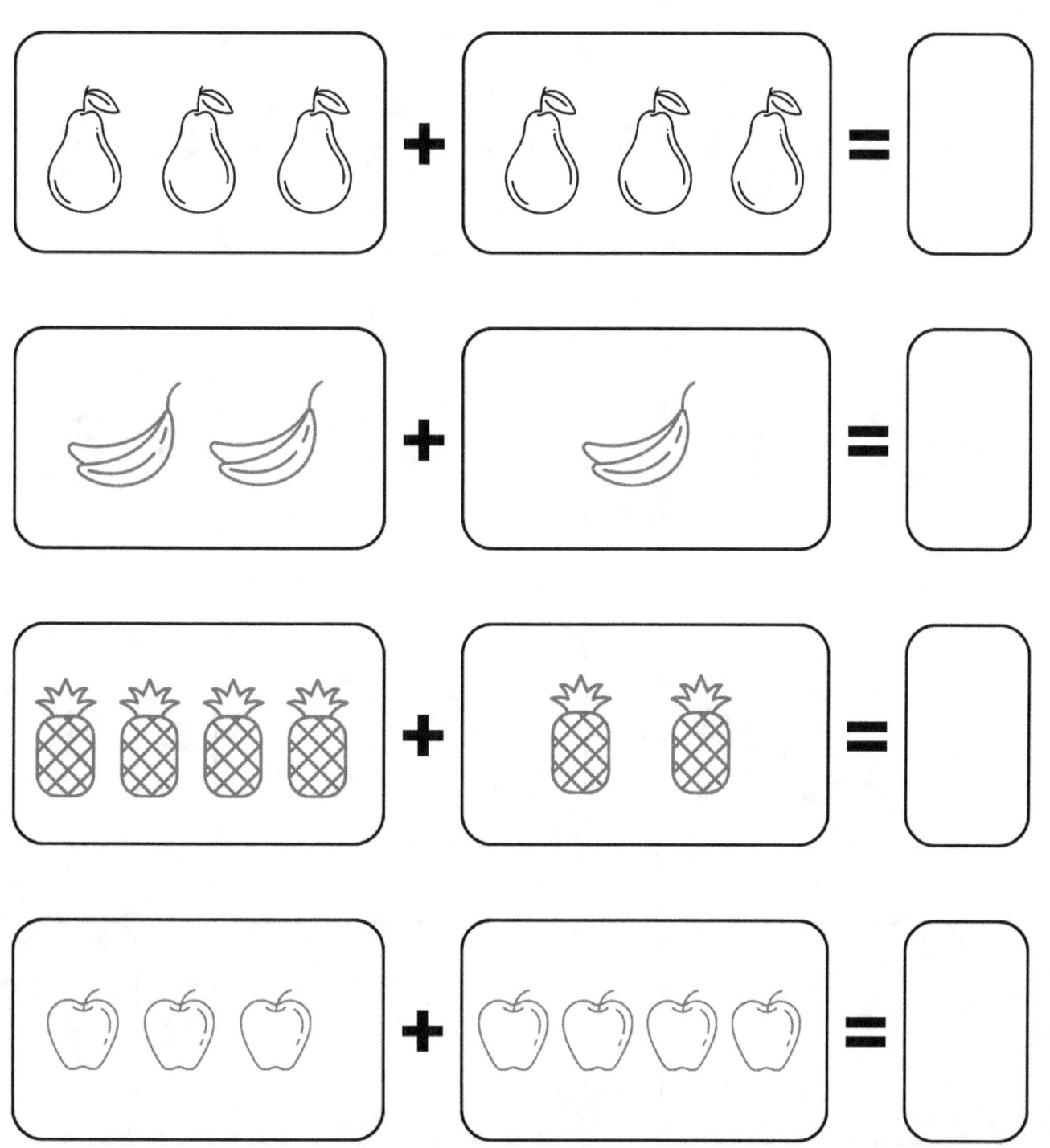

Congrats! You did it! Here's the answer for you to compare with what you got.

🍐🍐🍐 + 🍐🍐🍐 = **6**

🍌🍌 + 🍌 = **3**

🍍🍍🍍🍍 + 🍍🍍 = **6**

🍎🍎🍎 + 🍎🍎🍎🍎 = **7**

Count, add and write the sum in the box.

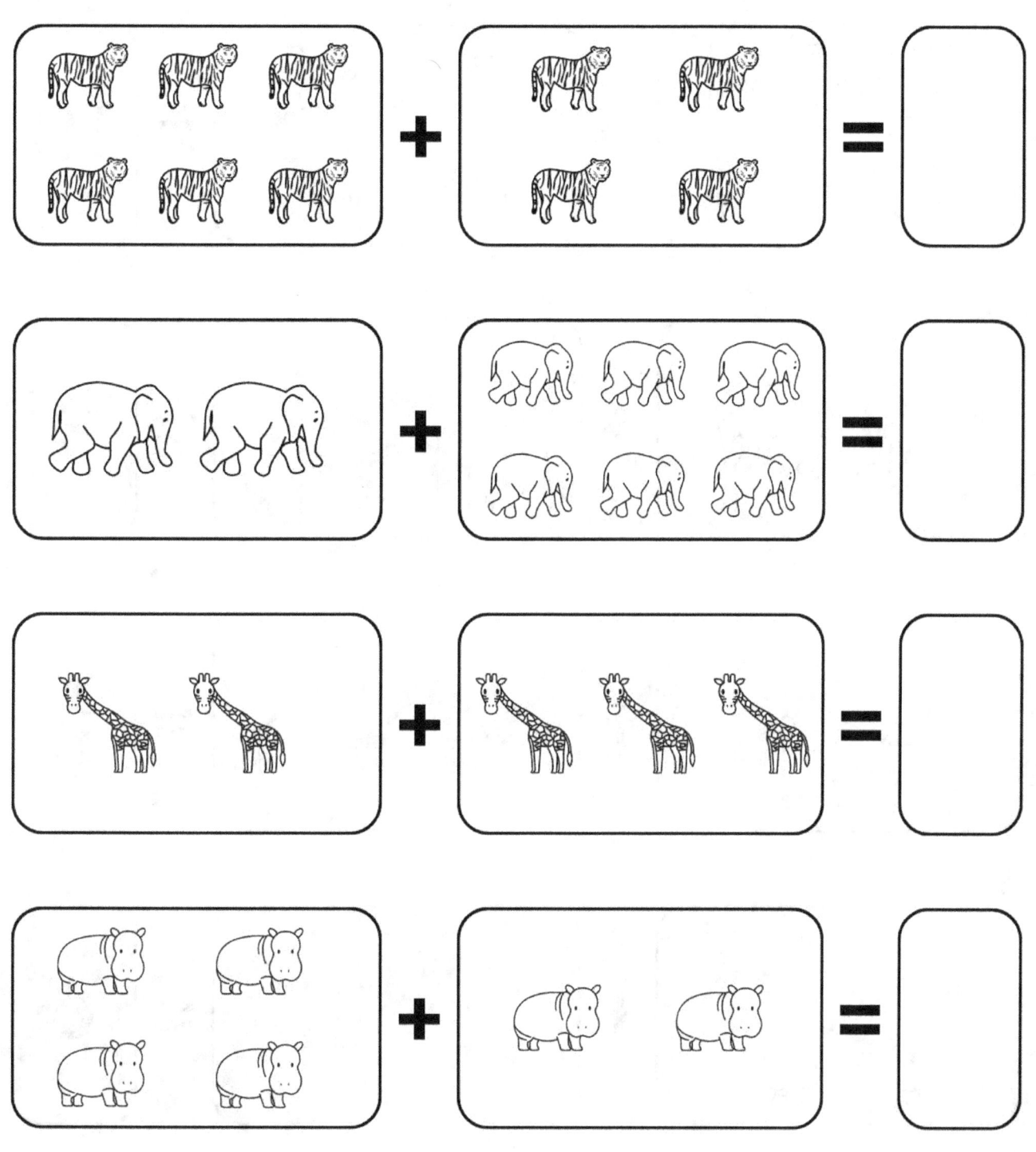

Congrats! You did it! Here's the answer for you to compare with what you got.

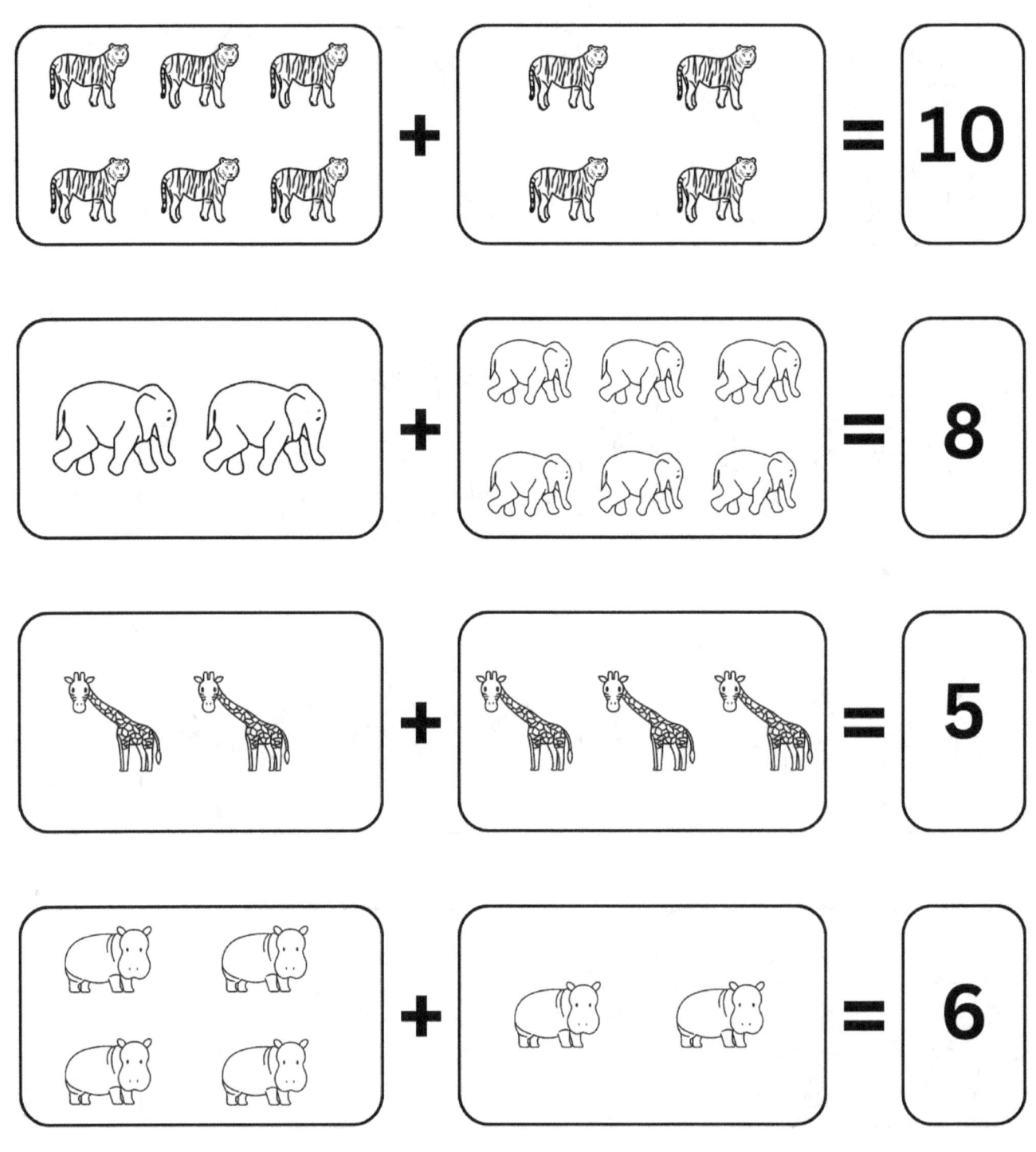

Count, add and write the sum in the box.

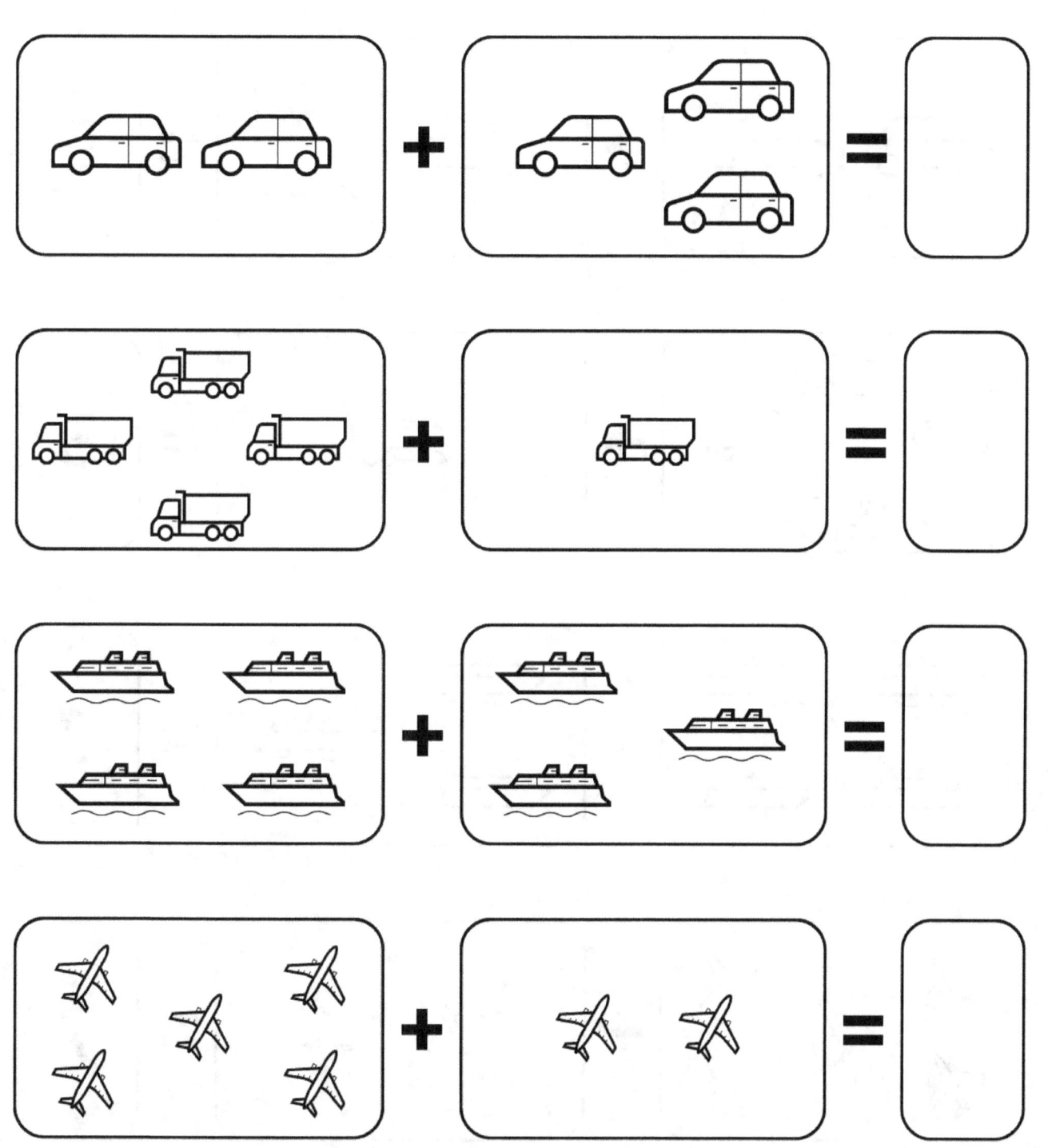

Congrats! You did it! Here's the answer for you to compare with what you got.

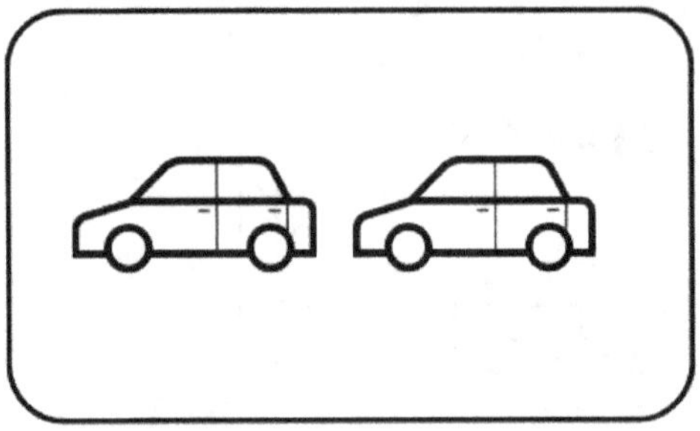

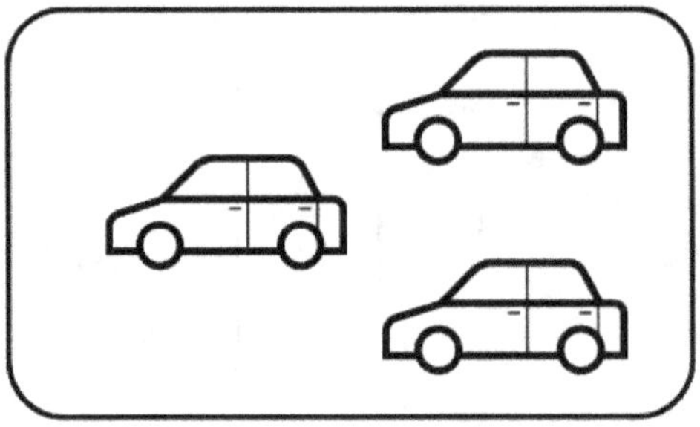

2 + 3 = 5

4 + 1 = 5

4 + 3 = 7

5 + 2 = 7

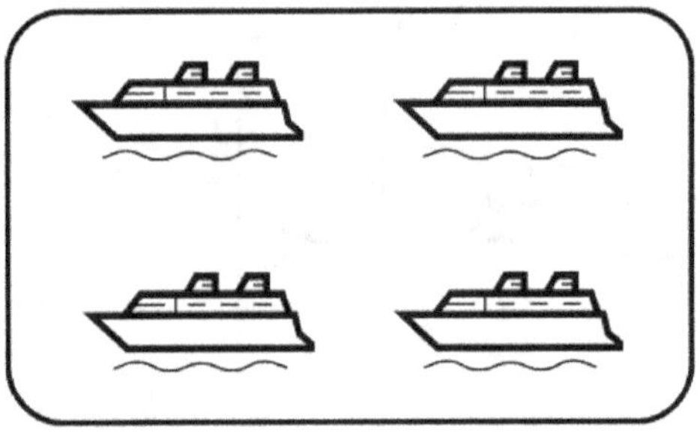

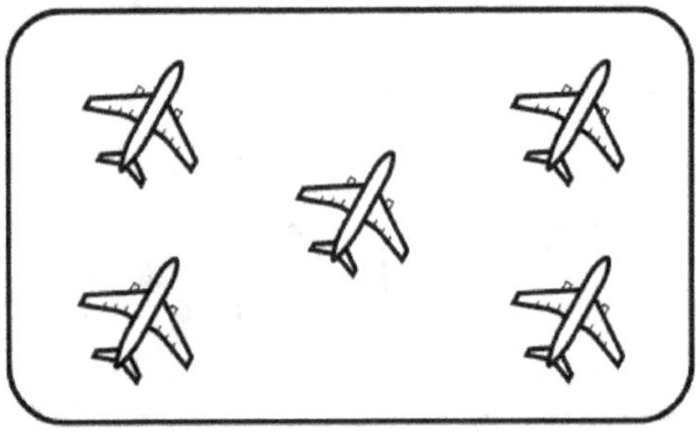

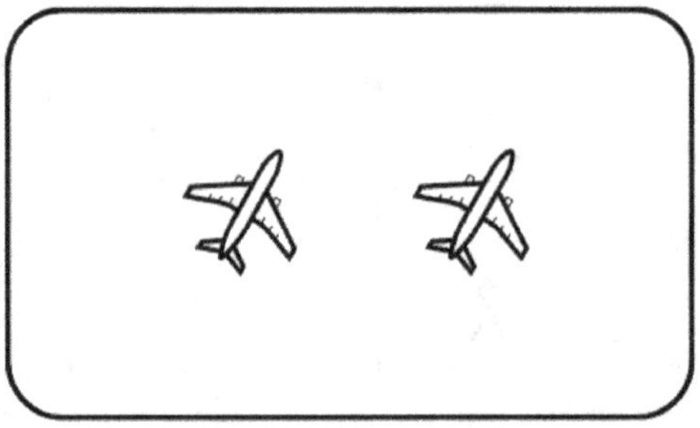

Get ready to connect the dots! Follow the numbers from 1 to 10 to reveal a surprise picture. Don't forget to color it in!

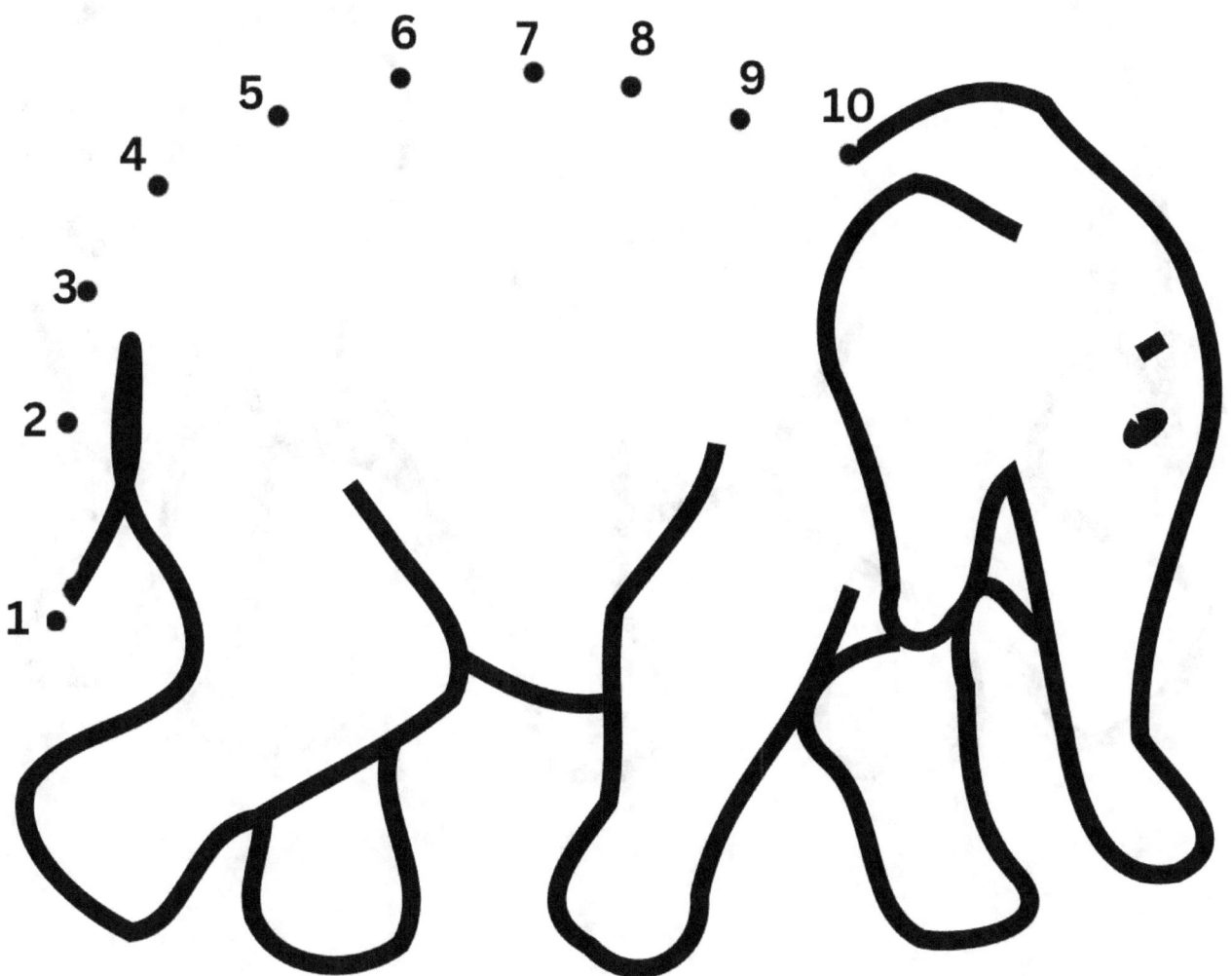

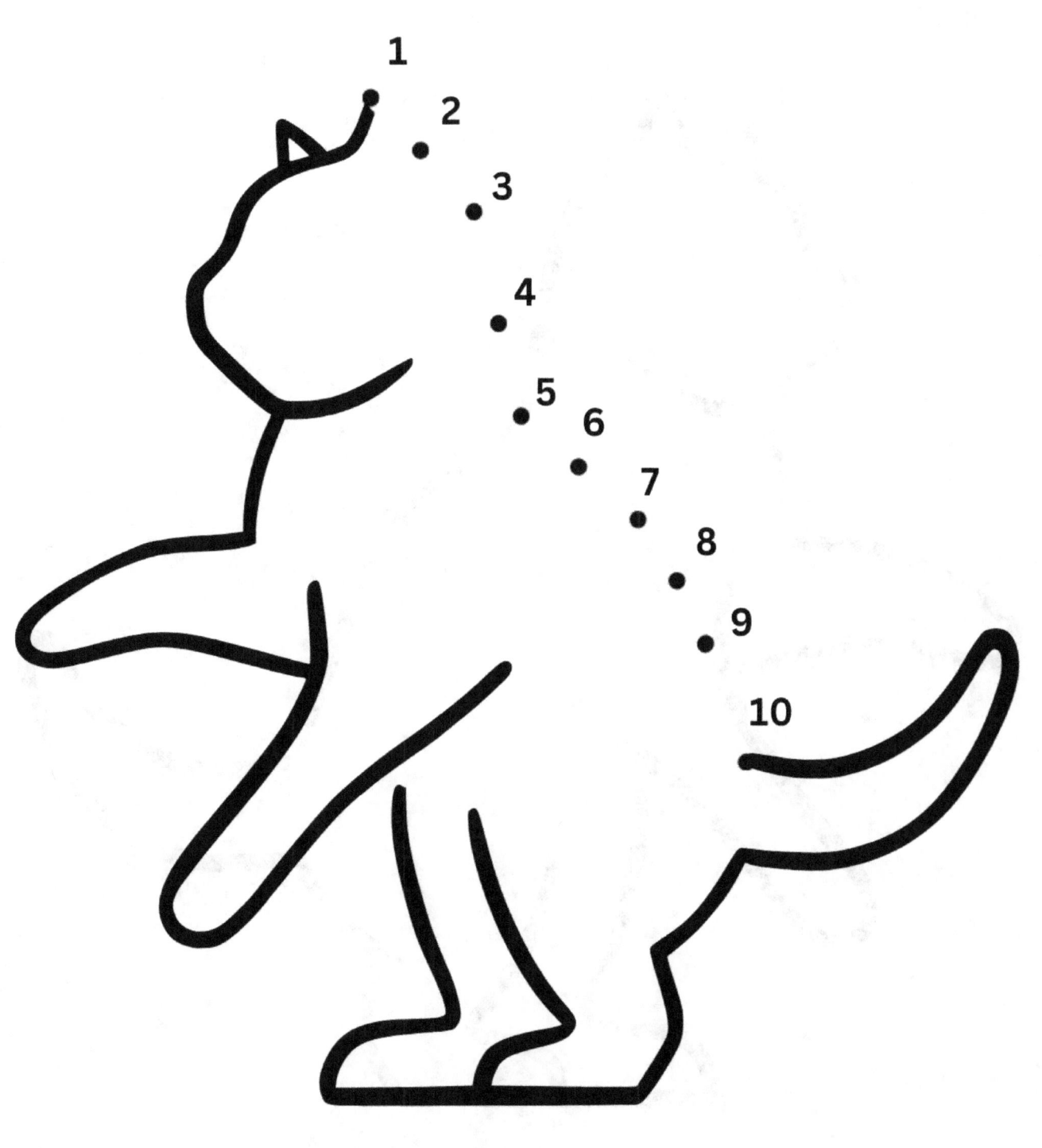

Cross out the correct amount and write the answer to see how many animals are left - let's go!

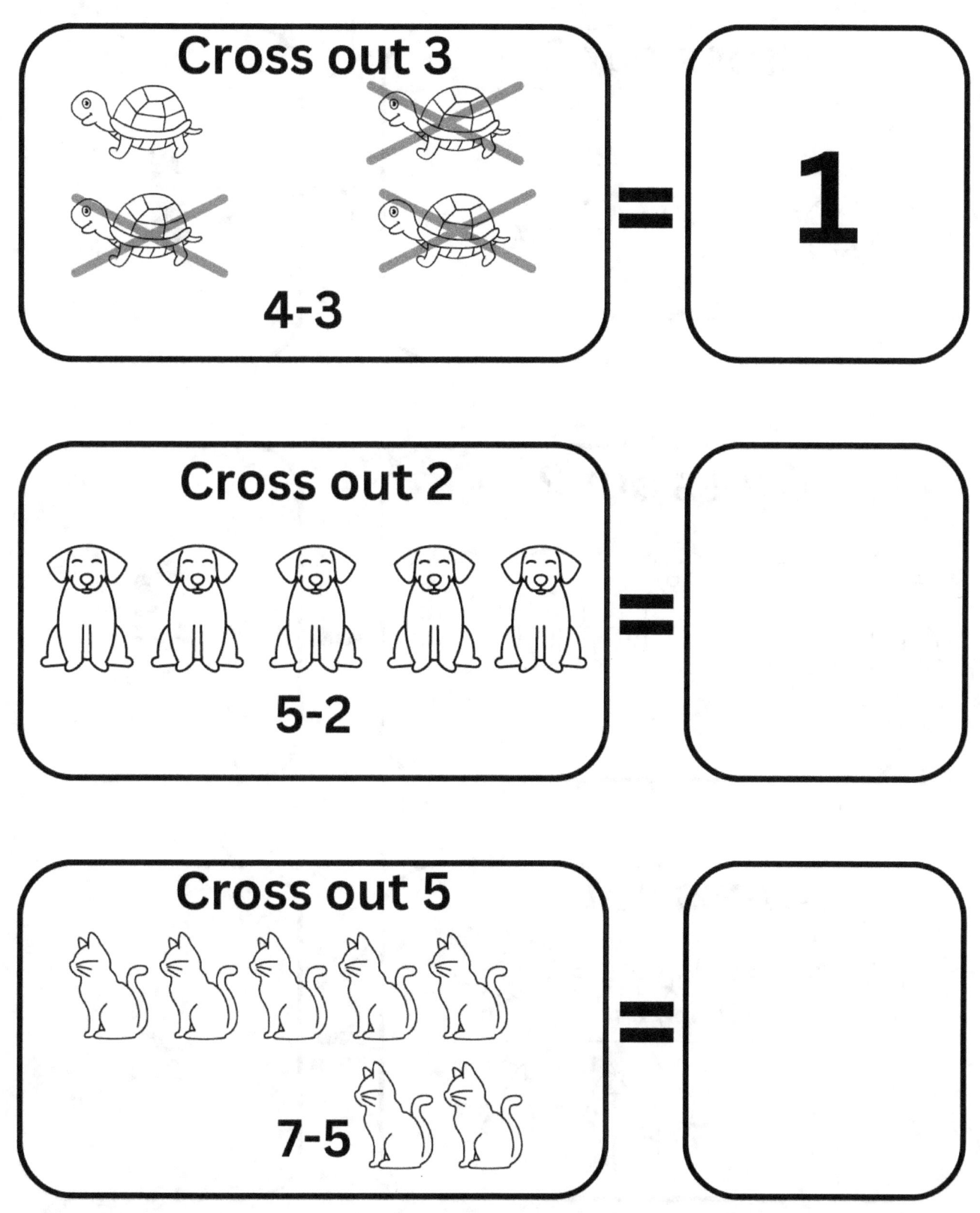

Cross out 3

4-3

= 1

Cross out 2

5-2

=

Cross out 5

7-5

=

Congrats! You did it! Here's the answer for you to compare with what you got.

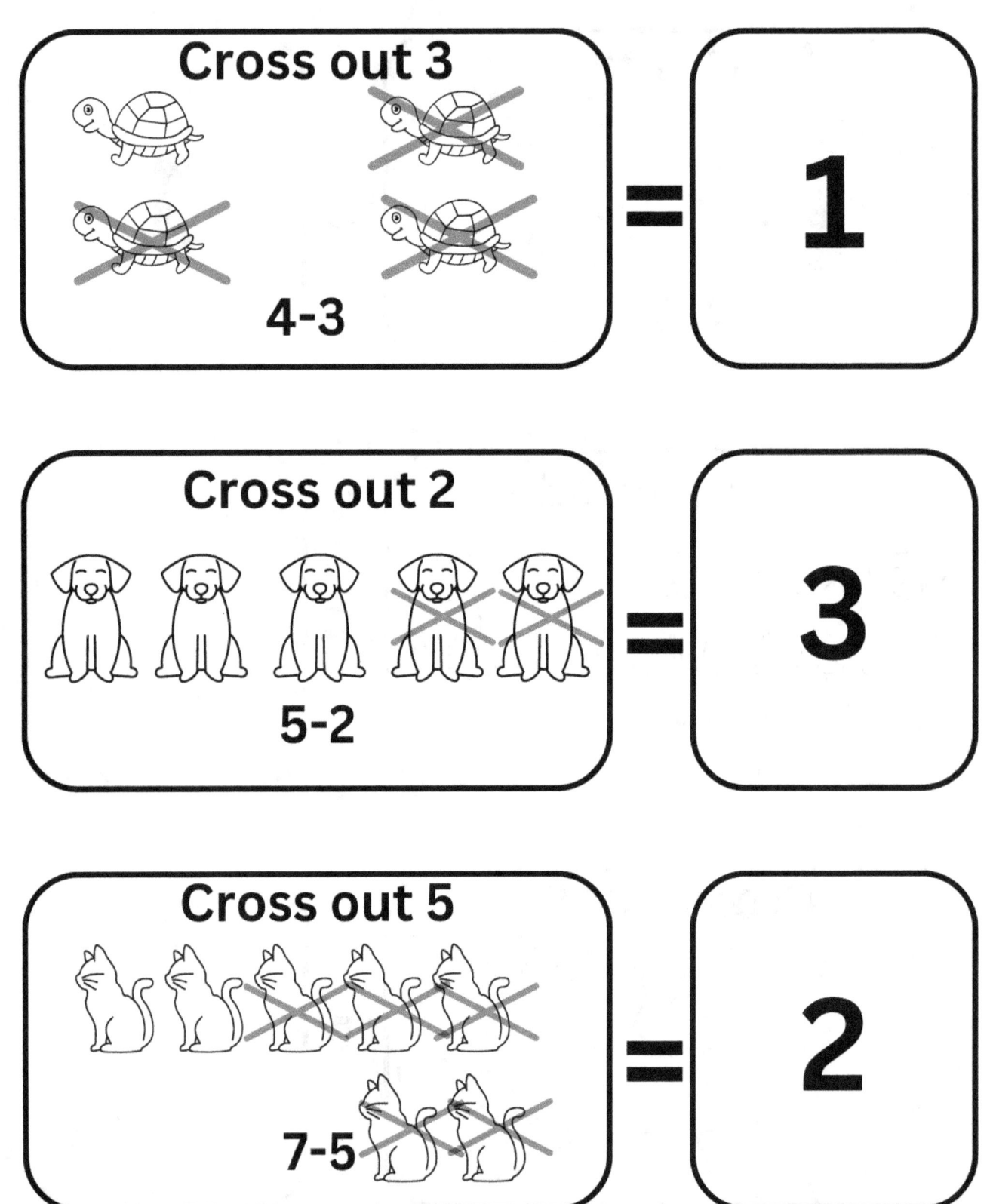

Cross out 3
4-3
= 1

Cross out 2
5-2
= 3

Cross out 5
7-5
= 2

Cross out the correct amount and write the answer to see how many animals are left - let's go!

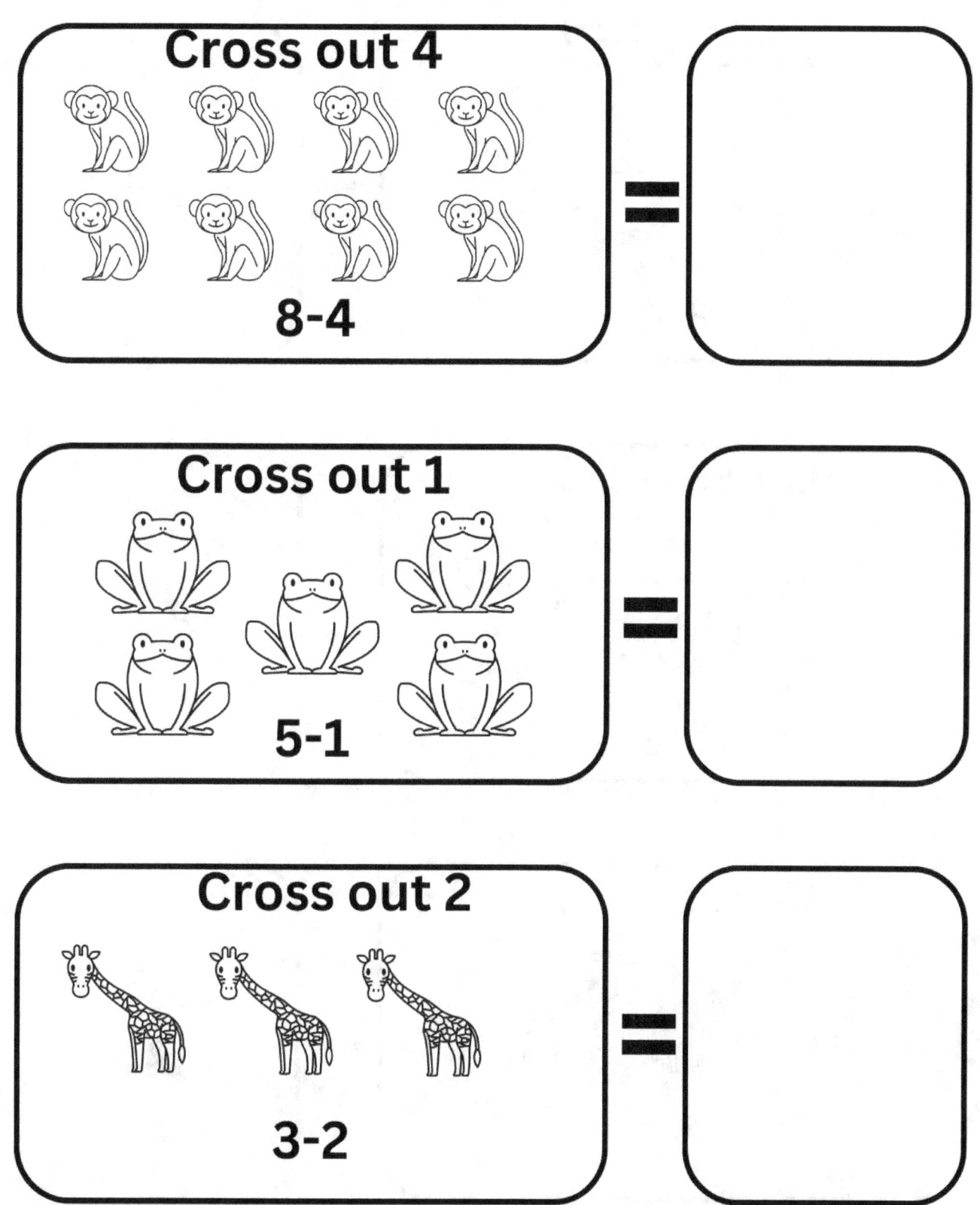

Cross out 4

8-4

=

Cross out 1

5-1

=

Cross out 2

3-2

=

Congrats! You did it! Here's the answer for you to compare with what you got.

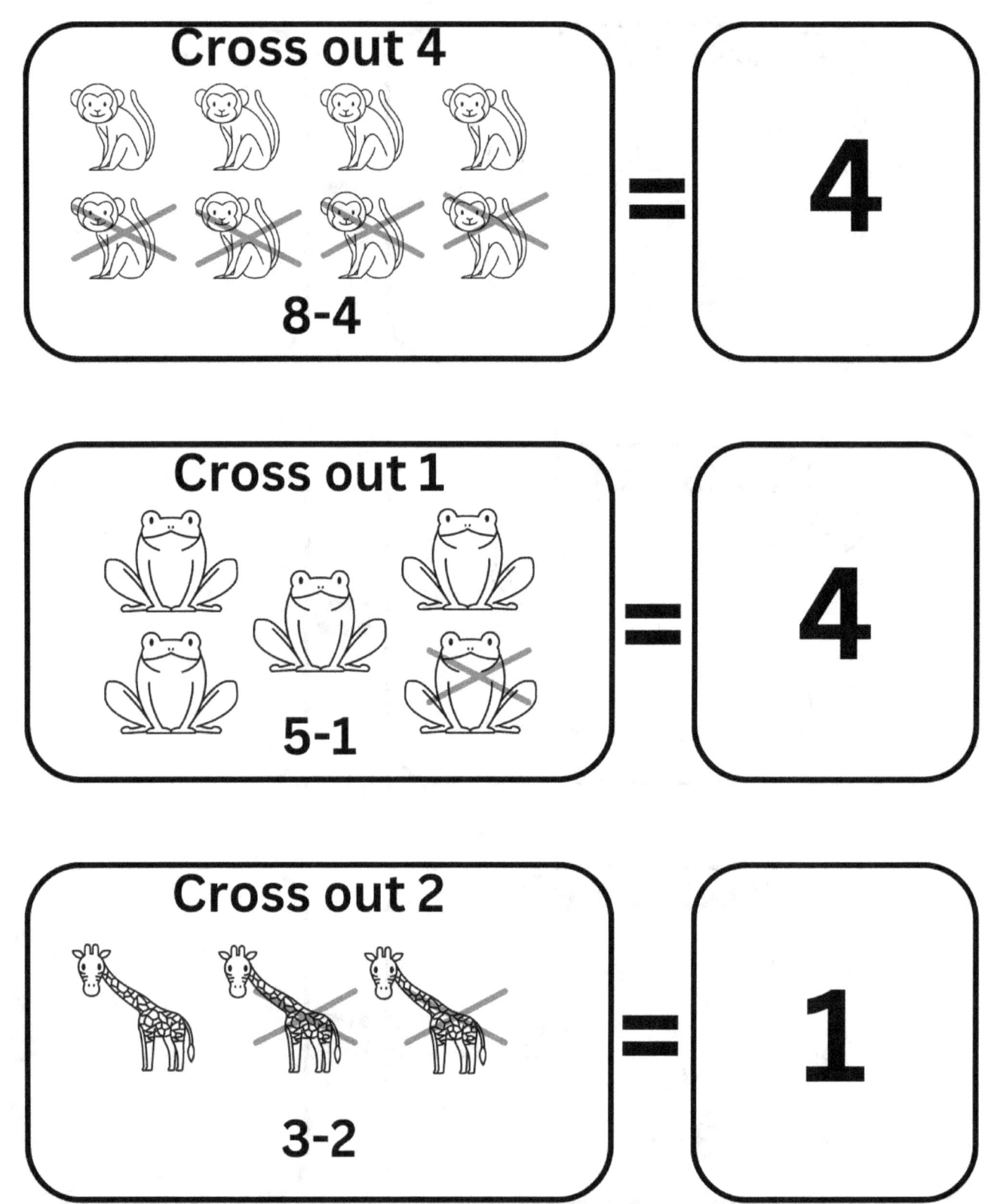

Cross out 4

8-4

= 4

Cross out 1

5-1

= 4

Cross out 2

3-2

= 1

Cross out the correct amount and write the answer to see how many animals are left - let's go!

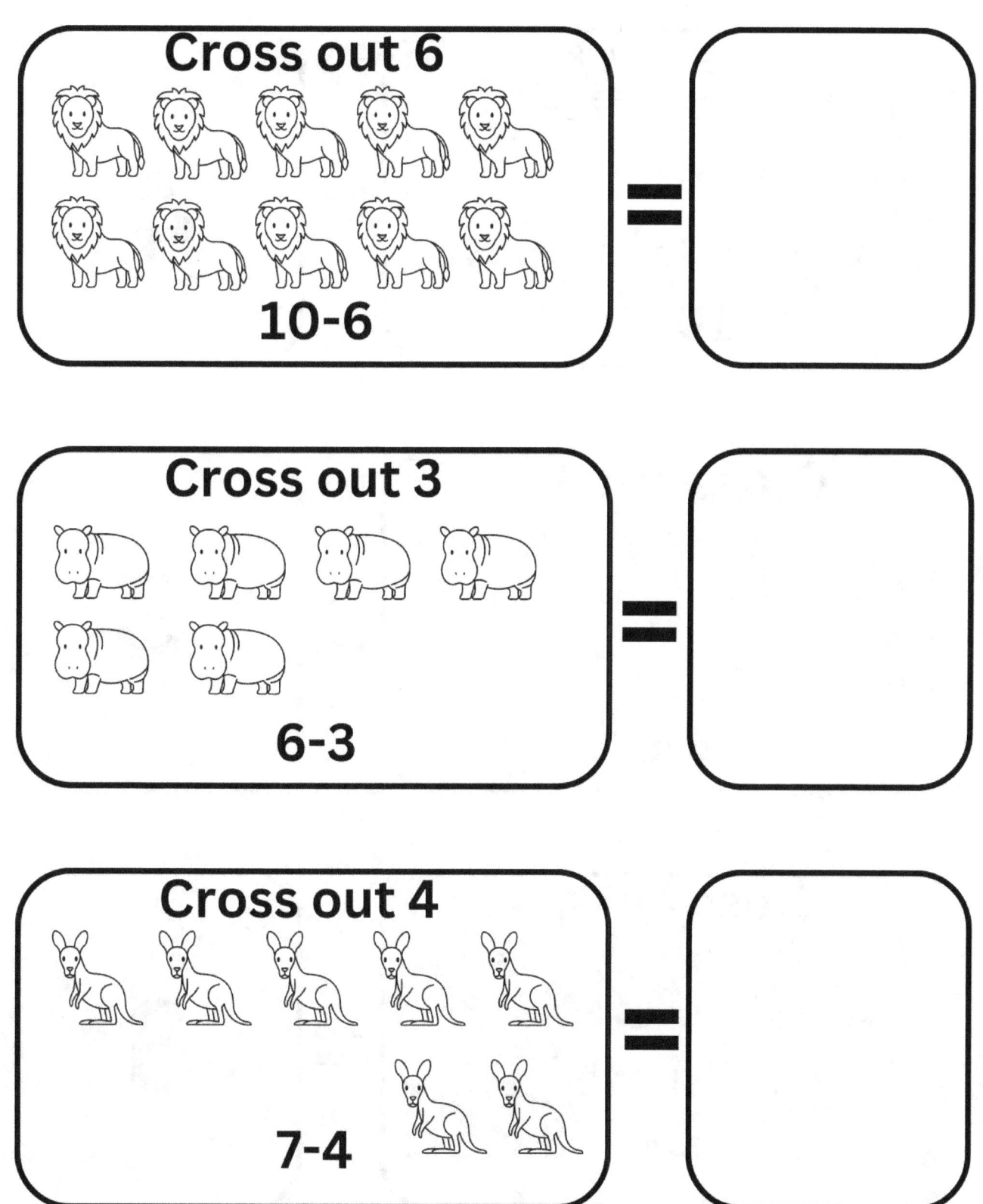

Cross out 6

10-6

=

Cross out 3

6-3

=

Cross out 4

7-4

=

Congrats! You did it! Here's the answer for you to compare with what you got.

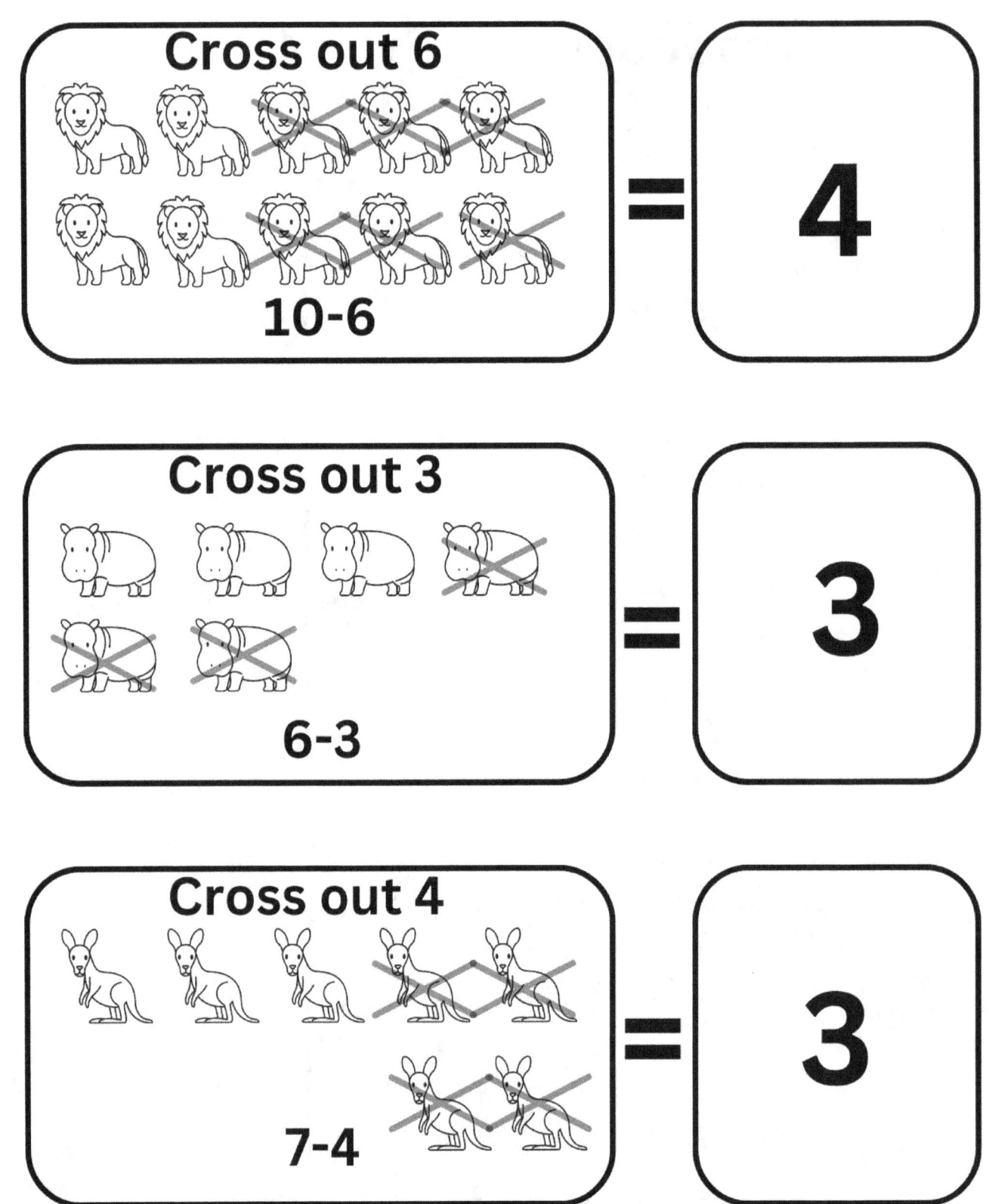

Cross out 6

10-6

= 4

Cross out 3

6-3

= 3

Cross out 4

7-4

= 3

Cross out the correct amount and write the answer to see how many animals are left - let's go!

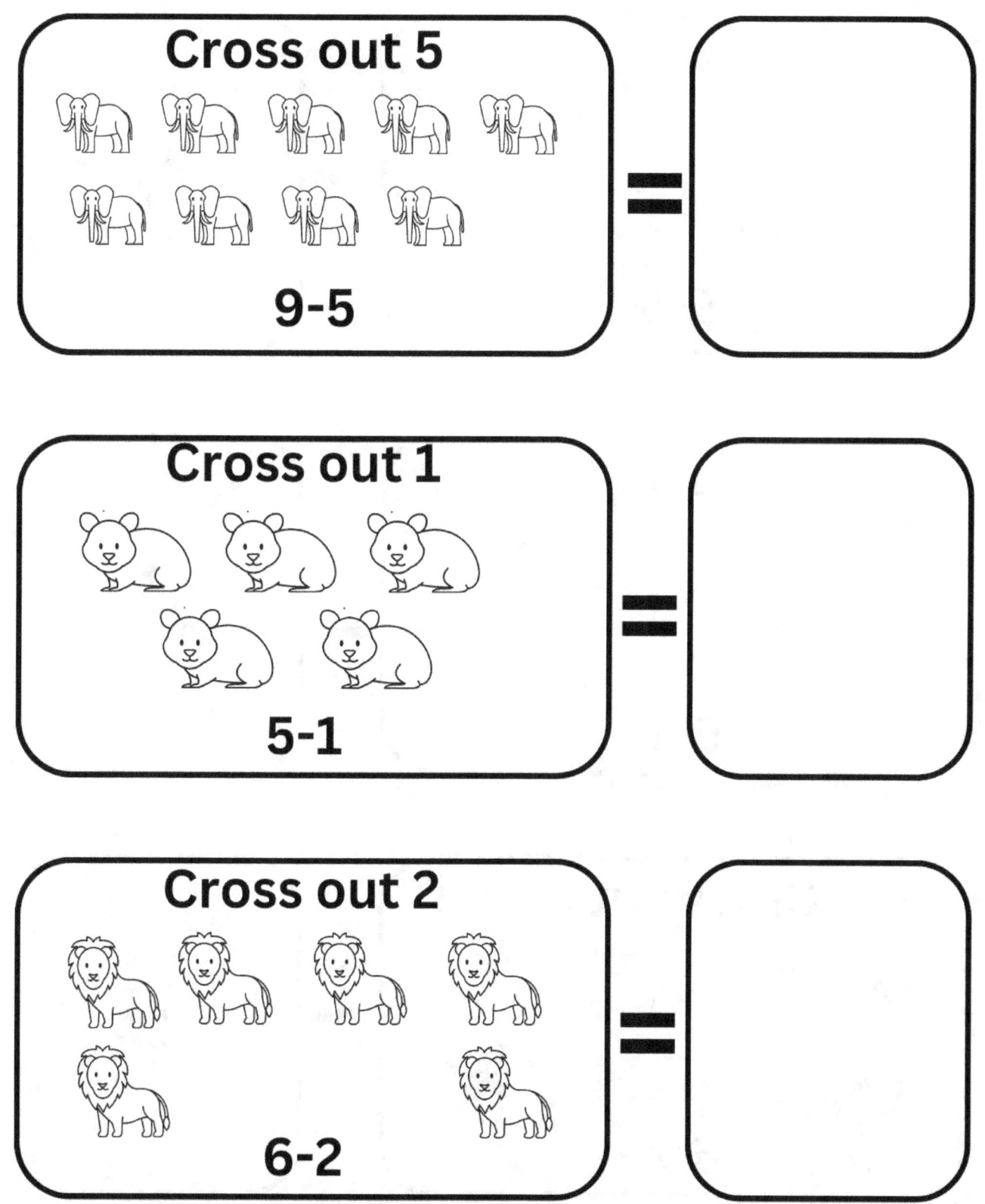

Cross out 5
9-5
=

Cross out 1
5-1
=

Cross out 2
6-2
=

Congrats! You did it! Here's the answer for you to compare with what you got.

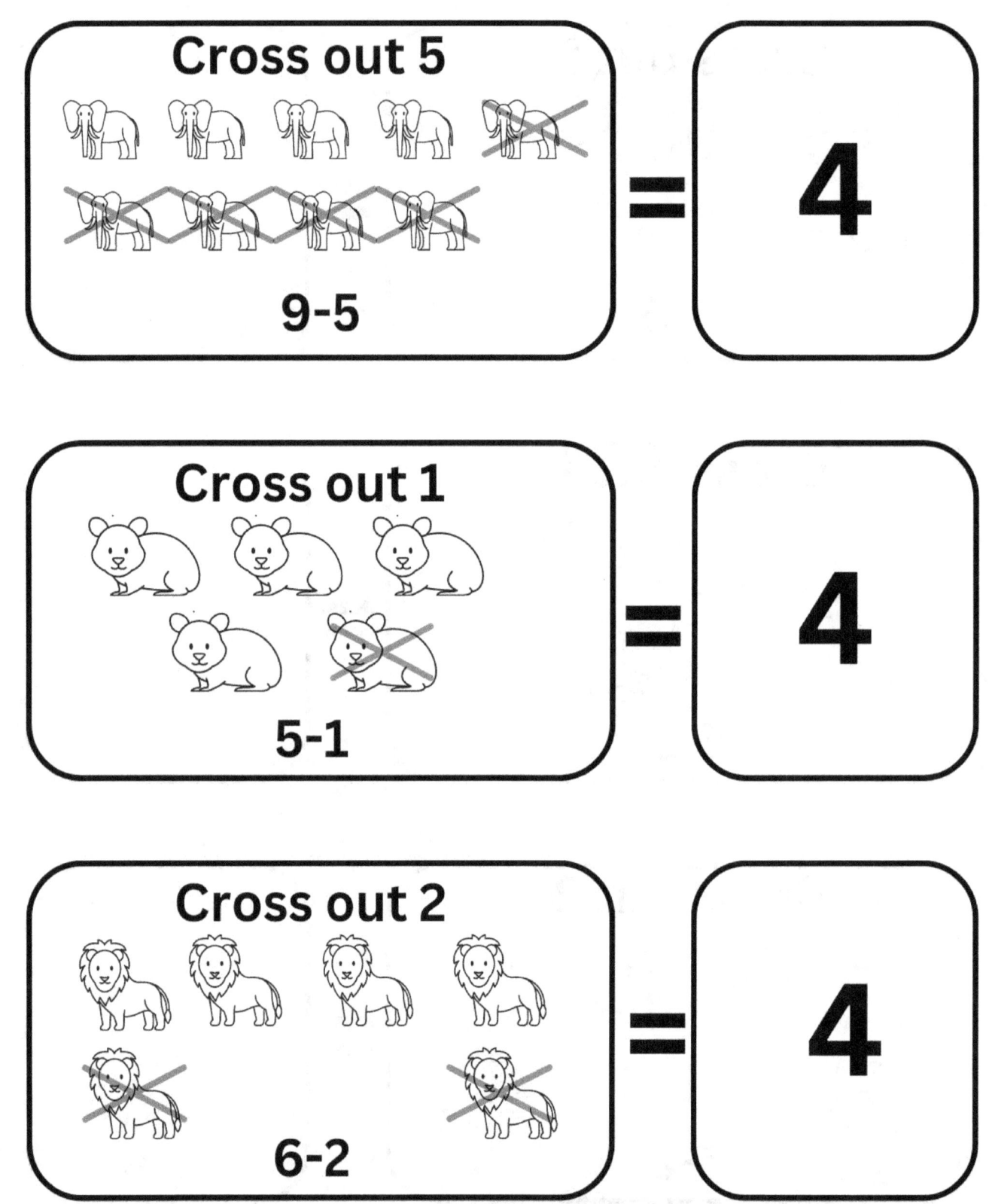

Cross out 5

9-5

= 4

Cross out 1

5-1

= 4

Cross out 2

6-2

= 4

Cross out the correct amount and write the answer to see how many animals are left - let's go!

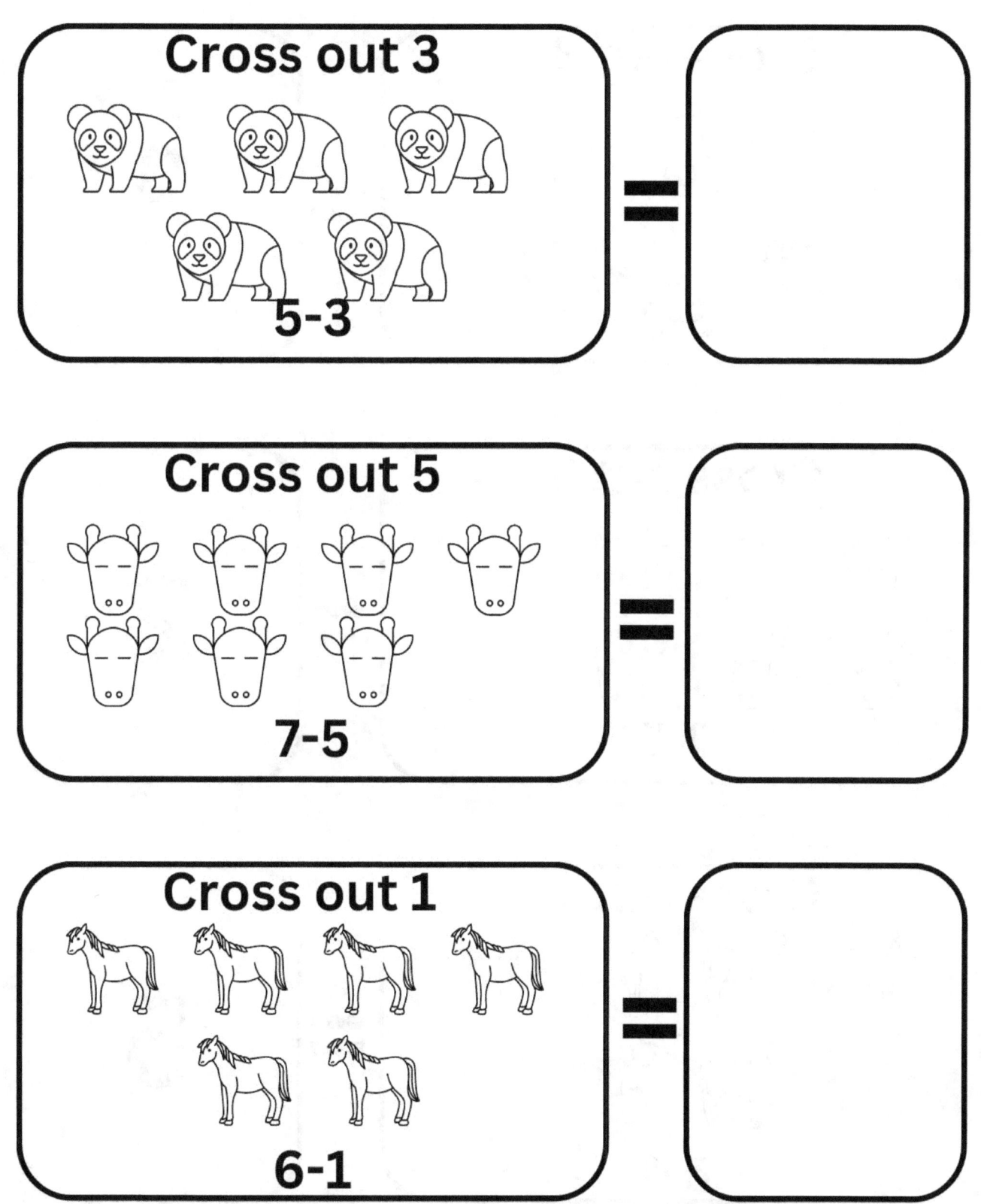

Cross out 3

5-3

=

Cross out 5

7-5

=

Cross out 1

6-1

=

Congrats! You did it! Here's the answer for you to compare with what you got.

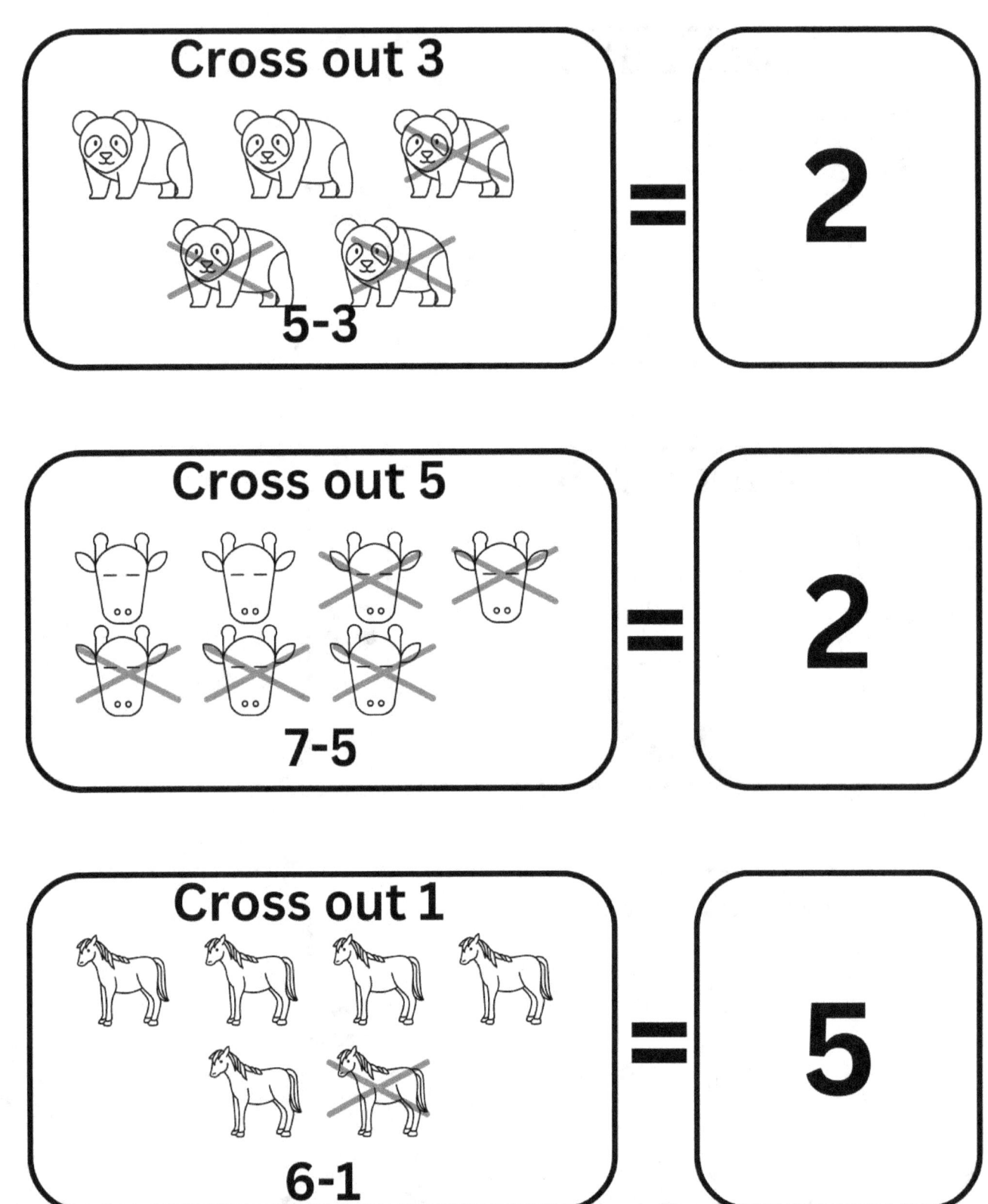

Cross out 3

5-3

= 2

Cross out 5

7-5

= 2

Cross out 1

6-1

= 5

Well done on mastering counting from 1 to 10! You've taken a big step in your math journey, and I'm so proud of you. Now, it's time to continue building your skills and learning more about numbers.

Our next book is designed to help you count even higher, all the way from 10 to 20. You'll get to review the numbers from 1 to 10 to make sure you still remember them, and then you'll learn how to count the numbers from 11 to 20.

So get ready to have some fun and take your math skills to the next level with our next book! Let's learn together and continue growing our love for math.

Would you like to receive some fun freebies for your child, such as dot-to-dot pictures and a math diploma? Simply send us an email with the subject line "Preschool Math 1-10" to claim your free resources!

preschoolmath101@gmail.com